AF335274

MODELING AND PARAMETER ESTIMATION IN RESPIRATORY CONTROL

MODELING AND PARAMETER ESTIMATION IN RESPIRATORY CONTROL

Edited by
Michael C. K. Khoo

School of Engineering
University of Southern California
Los Angeles, California

PLENUM PRESS • NEW YORK AND LONDON

Library of Congress Cataloging in Publication Data

Biomedical Simulations Resource Short Course on Modeling and Parameter Estimation
in Respiratory Control (4th: 1989: Marina del Rey, Calif.)
  Modeling and parameter estimation in respiratory control / edited by Michael C. K.
Khoo.
     p.   cm.
  "Proceedings of the Fourth Annual Biomedical Simulations Resource Short Course
on Modeling and Parameter Estimation in Respiratory Control, held May 21–22, 1989,
in Marina del Rey, California" — T.p. verso.
  Includes bibliographical references.
  ISBN 0-306-43530-6
  Includes index.
  1. Respiration — Regulation — Mathematical   models — Congresses.   2. Parameter
estimation — Congresses. I. Khoo, Michael C. K. II. Title.
  [DNLM: 1. Models, Biological — congresses. 2. Respiration — physiology — congresses. WF 102 B6155m 1989]
QP123.B56   1989
612.2 — dc20
DNLM/DLC   90-7301
for Library of Congress   CIP

Proceedings of the Fourth Annual Biomedical Simulations Resource Short Course
on Modeling and Parameter Estimation in Respiratory Control,
held May 21–22, 1989, in Marina del Rey, California

© 1989 Plenum Press, New York
A Division of Plenum Publishing Corporation
233 Spring Street, New York, N.Y. 10013

Printed in the United States of America

Dedicated to the Memory of
Fred S. Grodins
November 18, 1915 – September 21, 1989
Physiologist, Modeler, and Friend
"There is nothing magic about Models (or is there?)!"
—FSG, 1981

PREFACE

Experimentalists tend to revel in the complexity and multidimensionality of biological processes. Modelers, on the other hand, generally look towards parsimony as a guiding principle in their approach to understanding physiological systems. It is therefore not surprising that a substantial degree of miscommunication and misunderstanding still exists between the two groups of truth–seekers. However, there have been numerous instances in physiology where the marriage of mathematical modeling and experimentation has led to powerful insights into the mechanisms being studied. Respiratory control represents one area in which this kind of cross–pollination has proven particularly fruitful. While earlier modeling efforts were directed primarily at the chemical control of ventilation, more recent studies have extended the scope of modeling to include the neural and mechanical aspects pertinent to respiratory control. As well, there has been a greater awareness of the need to incorporate interactions with other organ systems. Nevertheless, it is necessary from time to time to remind experimentalists of the existence of modelers, and vice versa. The 4th Annual Biomedical Simulations Resource (BMSR) Short Course was held in Marina Del Rey on May 21–22, 1989, to acquaint respiratory physiologists and clinical researchers with state–of–the–art methodologies in mathematical modeling, experiment design and data analysis, as well as to provide an opportunity for experimentalists to challenge modelers with their more recent findings. Four areas, in which there has been significant modeling activity in the past several years, were chosen as sub–themes for this forum: respiratory control during exercise, empirical models and parameter estimation, control of breathing during sleep, and neural control of respiration. This book is based on the presentations delivered at that meeting, and its chapters are subdivided according to the above four sub–themes.

I wish to acknowledge with deep gratitude the financial support provided by our sponsor, the Biomedical Research Technology Program, Division of Research Resources of the National Institutes of Health. The arrangements for the meeting could not have been organized any better, thanks to the unfailing dedication and efficiency of Mrs. Gabriele Larmon and her staff. In order to maintain a high level of uniformity in the appearance of the several chapters of this book, many manuscripts were retyped and re–retyped; I am indebted to Mrs. Diane Lord for her patience and proficiency in word–processing. I am grateful for the advice and encouragement given by Dr. Vasilis Marmarelis and Dr. David D'Argenio of the BMSR, and Dr. H.K. Chang, Chairman of the Biomedical Engineering Department of USC. Special thanks go to Dr. Stanley Yamashiro for his support and help in chairing one of the sessions of the meeting. Finally, I must pay special tribute to the late Professor Fred Grodins: through his pioneering works, he laid much of the foundation for modern respiratory control modeling. His indomitable wit and intellect will be greatly missed by those of us who have known him personally or through his writings.

Michael C.K. Khoo

Los Angeles
October 1989

CONTENTS

Part I:

MODELING OF RESPIRATORY CONTROL DURING EXERCISE

WHY AND HOW ONE MODELS EXERCISE ON A COMPUTER (A Tutorial)

William S. Yamamoto

Department of Computer Medicine
George Washington University, Washington, D.C.

WHY

The principal reason for modeling exercise or any physiological phenomenon is simply stated, "because we do not understand all we know." This is particularly true in the mature parts of science where the principal characteristics of the subject have been described in a corpus of literature encompassing conceptually interlocking experiments and where conflicting or partial theory exists. Theory which is the formulation of a causal myth to explain elements in a catalog of observation serves the function of economizing on memory and increasing human control over processes that the knowledge describes, i.e., technology. Modeling, then, is no more than publication in an idiom which recalls many considerations necessary to any given situation and discovers new or unexpected implications of old systems of ideas. Instead of dialectic, computer simulation with models produces unambiguously the implied consequences of the assumptions which include both the declarative and the procedural forms of knowledge. Modeling is parsimonious. In physiology the variety of laboratory experiments that pertain to a global function like exercise is redundant exposition of a common underlying mechanism.

Effective modeling is not phenomenological but theoretical. The test of a successful model is in the simulation. One is thus driven ineluctably to the view that there are probably several theories adequate to explain a given set of phenomena. However, it is an unstated article of faith in science that there is one "true" mechanism that explains all the phenomena and therein lies a basic issue in the philosophy of science. For design and technology, partial and even "incorrect" models are sought and used. These are effective in reproducing or controlling some subset of the phenomena that a physiological system can produce.

Models are made for two major reasons: first, to explain, by constructing, a common mechanism for all of a large set of phenomena. Second, to design machines or procedures to control technologically the organism (as in medical care or extending environmental capability) in a practical way.

HOW

Verbal models are familiar to all of us in the form of scholarly papers and monographs. They are simulated in the mind and evaluated by dialectic. They are written in a natural language. This experience defines for us how computer models are to be made. One first selects a domain of discourse – in the present case, the exercising mammal, and a language.

The language one uses places limitations on the domain of discourse. For exercise and most physiological processes that are continuous in time, mathematical equations in any procedural language is an effective choice. It is my prejudice that the language of the model should be of a low level and permit the symbolic mapping of many variables. FORTRAN remains a good choice. For the most part, one models processes with entities or attributes that have quantitative measure. Since useful models must speak to an audience that characteristically uses verbal models, mathematical computer models should be structured so that as many magnitudes as possible have verbal equivalents. This leads to inefficient algorithms. Efficient lumped parameter models often result in explanations that use variables which have no familiar biological connotation. On the other hand, lumping systemic entities like blood volume or tissue are effective. The result is that the largest class of useful models are systems of differential and algebraic equations to which are attached appropriate glossaries, reference materials, and input output systems. From such lumping, model components traverse the levels of discourse common in biological reasoning, starting with the whole organism as an autonomous entity, and addressing successively systemic, organ, tissue, cell, organelle, and biochemical quantities. Since exercise involves nearly every body system in the short run and all in the long run, the completeness of the model is arbitrary and depends on the unstated intensity of coupling between body parts. It may seem most desirable to write models using only the most elementary physical process. Such models, however, are not only computationally difficult and inefficient but lack the capacity to simulate the phenomenon of emergent properties. The original requirement for mapping colloquial dialectic with computer process fails.

In a problem like exercise, models are double and include models for the exercising system and the forcing system; i.e., one part is the organism that performs exercise and the other its environment, e.g., in exercise, the onset, duration, severity, cessation, or inspired gases.

Where possible, physical laws or colloquial description of observed laboratory entities are described mathematically. Thus, material flow models have immediate familiarity. The principle of conservation of material applies if one allows for biochemical transformation. In respiratory exchange the stoichiometric relationships are simple and signals have chemical or physical form. Effective models of exercise thus contain the vocabulary of physiology. Some system components like the respiratory complex of the brain stem cannot be stated on experimental grounds. To be executable, fictional hypotheses must be incorporated. Model simulation then tests the whole (both "truth" and "theory") as a quest for emergent properties which are observables. For example, the hyperpnea of exercise emerges from the action of model components but is itself not explicitly in the model.

There is yet a third facet to the model. Computer algorithms stand apart; they are not statement of physical law, biological observables, nor biological hypotheses. Algorithms concern differential equation solvers, algebraic constants and equations, data structures including the simplest array, stacks, and trees, and some relational calculus, things that cause hardware to "understand" the symbols. Similar computational methods permit the use of models by coupling them to laboratory or clinical research. Such use requires entry in the whole domain of signal analysis, and engineering instrumentation.

Many models have been written to study pulmonary ventilation in exercise. The field may be the most active of all in physiological modeling. One can no longer provide a terse bibliographic listing – but thanks to automated library cataloguing systems, a student does not lack for entree into this study. One should, however, draw attention to the triennial meetings and the attendant volumes that started in Oxford in 1978[1,2]. I shall try now to illustrate the argument primarily with things I have done.

The computer model as a construct for codifying knowledge is shown in Fig. 1. It forms an active linkage[3,4,5] between the human activities in the laboratory and technology and his efforts to deduce implications and codify the aggregate of human experience.

There must be some morphological (and therefore linguistic) continuity between a the-
matical model and the organism. In the case of exercise, the content of the model is drawn
from the systemic view of organism, selecting only subsystems that are conceptually separa-
ble but coupled together and believed to be necessary participants in exercise. Circulation,
pulmonary ventilation, a neural mechanism, and a metabolism seem inextricably united. A
choice in the level of detail for each is required[5] (Fig.2). The block diagram illustrates the
intention to include neural and humoral control and an extended spatial character for the
body. It also portrays multiple levels of detail, i.e., ventilation is made up of breaths, blood
volume out of aliquots, and tissues out of micro circulatory beds. Thus several levels of discourse are represented. The tissues are visualized lumped at a microscopic scale into three
compartments.

The literature is searched for descriptive data which can be formulated for each part.
Where there is disagreement in the literature, one is forced to choose between sources. Choice
from the literature should be representable and authenticated and fictional constructs should
be plausible. This choice, done privately, is the dialectic of the scientific meeting place.

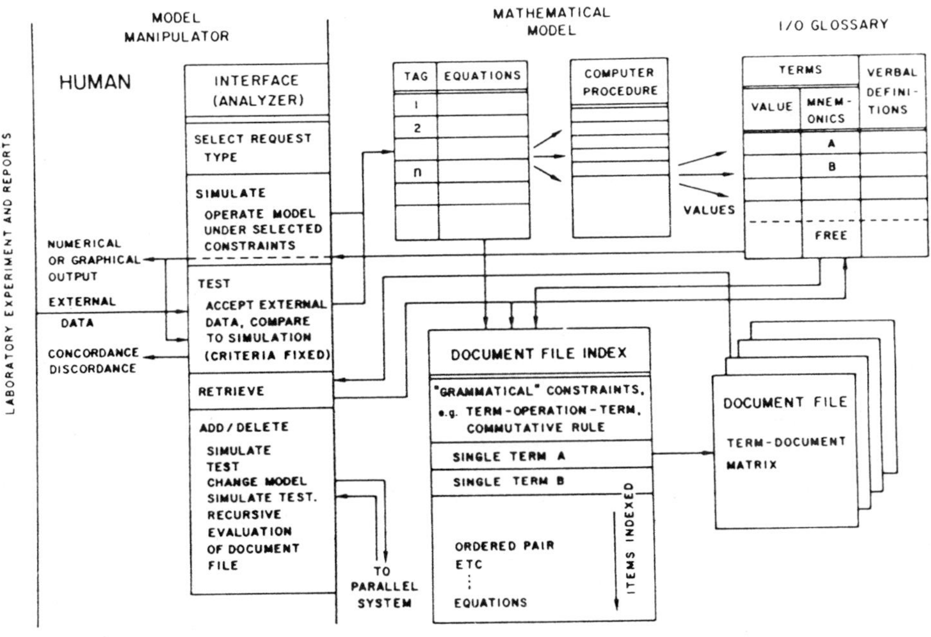

Figure 1.   Schematic of the principal relationships between a model, its
documentary basis, and interface to experiment[3].

Where there is no laboratory or clinical observation, one writes hypothetical equations.
Therein is expressed the "scientific creativity" in modeling. All choices are guided by the
relevance of the choice to the emergent behavior one seeks to simulate.

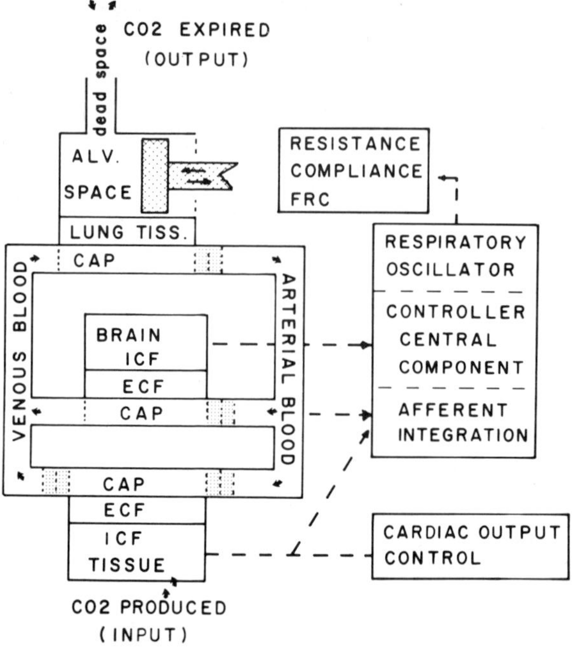

Figure 2.   Block diagram of principal components of a complete model for regulation of CO_2 in a man[7].

Finally, one must specify performance criteria for the model (Fig.3). In the present instance[6], the criteria are the qualitative differences between the relation of arterial PCO_2 to hyperpnea in exercise and in CO_2 inhalation. We require the model to show isocapnic hyperpnea within 2 torr PCO_2 for simulated exercise between rest and an 8–fold increase and hypercapnic hyperpnea in response to 0–9% CO_2 inhalation which falls between slopes 2.0 and 3.7 as observed in man.

Since the performance is stated in terms of CO_2 vs ventilation, the model focus is on the organism's handling of this compound. Most certainly the organism is redundant and there may be several sufficient mechanisms for the hyperpnea of exercise. However, only one of these, CO_2 exchange, also addresses the CO_2 inhalation data – and it is this issue that has focussed the problem of the hyperpnea of exercise for nearly a century.

In Fig. 4, I demonstrate[6] that I have solved the problem of the hyperpnea of exercise. There are undoubtedly other solutions to explain exercise, and perhaps some models will not involve CO_2 and yet control CO_2, but the purpose of this model, to explain the hyperpnea of exercise, is accomplished. Credibility of the solution rests on two issues: (1) How plausible are the "mythical" equations for control and body function used? and (2) How extensible (or useful) is the model in expanding its simulation repertoire? its domain of explanation?

Let us address each of these briefly. The principal arbitrary equations in the model are those necessary to cause the pulmonary mechanism to move in tidal fashion, to adjust the cardiac output to the level of exercise metabolism, and to relate the composition of the blood to the controller. The controller[6] requires that the respiratory center have blood flow sensitive to CO_2, have intra– and extracellular compartments, and an intrinsic metabolism that produces CO_2 proportioned to its activity. The controller also has a central and peripheral component separated by a finite time lag. The controller equation is an expression which increases proportionally to the extracellular and intracellular PCO_2 but with different sensitivities – this is equivalent to a proportional and cellular gradient controller. There is an arbitrary constant, which is equivalent to embedding a Lloyd–Cunningham expression. A last term of the controller is a covariance (squared magnitude) of the difference in PCO_2 of the blood at the brain and at a peripheral site. This embodies what has been called the oscillation signal, but merely gives the controller a non–flat characteristic over the spectrum. Another facet of the controller requires assumption relating the quantity of activity generated neuronal CO_2 production. The positive feedback thus implied is not in the controller equation. Another way to look at the controller is that it is a polynomial of degree 2 in 2 variables intra and extracellular PCO_2. The components are plausible and physically motivated. Whether one accepts this explanation or not physiologically, the model proves by construction a set of relationships that simulate the exercise – inhalation problem ... both a why and a how.

The second question speaks to validation of a model by extending its simulation to behavior outside its formulation or to extend the formulation without changing its behavior. Several directions to go are obvious: hypometabolism and apnea, the role of a neural signal, the dynamics of transitions, role of oxygen. These also challenge the necessity of the control expression we used. While it is sufficient, it may not be necessary. For example, a neural signal proportioned to metabolism appears also to be sufficient. After modeling exercise (not inhalation) on a neural basis, signal interaction must be defined. How does a model incorporate redundancy? Sherrington a century ago provided the idea of occlusion as a form of neural signal interaction. Figure 6 incorporates[1] this concept as a synaptic connection. In the sensory system such connection is seen in lateral inhibition. Is there such in the respiratory center? The model now behaves the same under humoral drive alone, neural drive alone, or both (provided the metabolic equivalent of exercise is present.) In attempting to extend the definition to hypometabolic regimes[8,9], apnea fails to appear without specific manipulation of neural drive. Apnea does not follow hypocapnia[4] automatically (Fig.5). In hypometabolic states with and without isocapnia, the neural and humoral signals interact differently than the simple occlusion relationships of exercise.

Pursuit of a neural signal and signal interaction requires a better "myth" for neuronal structure of the respiratory oscillator. Representation of neural types as identified[10] by stimulation, ablation, and recording and having the necessary properties to imbed in an exercise model is a prodigious undertaking. Much work is being done along these lines using nonlinear differential equation systems. In fact, this constitutes a major focus of research in the field[1], particularly that based on the single Van der Pol equation. We tried to represent neuron groups as ecology sets of Volterra equations in a non–linear space (i.e., with no multiplication in a non–negative domain). The former dissociates the morphological substrate from the equations and the latter run into an obscenely complex combinatorial problem. In either case, the modeling of neural signal in exercise must be deferred until the neural models show sufficient verisimilitude for incorporation into an exercise model.

Another validation of a model is to extend the domain of definition of a model by simulating exercise phenomena other than the test formulation. For example, exercise performed with various volumes of artificial external dead space has been reported[11]. With no model change except to change the numerical magnitude of dead spaces, the experimental data can be readily simulated (Fig.7)[4]. No new assumptions are necessary to imitate the body's adaptation to external dead space.

Small details lead to unexpected inferences. The small discontinuity in the derivative of the CO_2 inhalation curve[6] (Fig.4) was found to depend upon the describing function for cerebral blood flow. A piece–wise continuous CO_2 vs. blood flow curve, a linear curve, and a logistic curve have been reported in the literature. When scaled appropriately the exercise

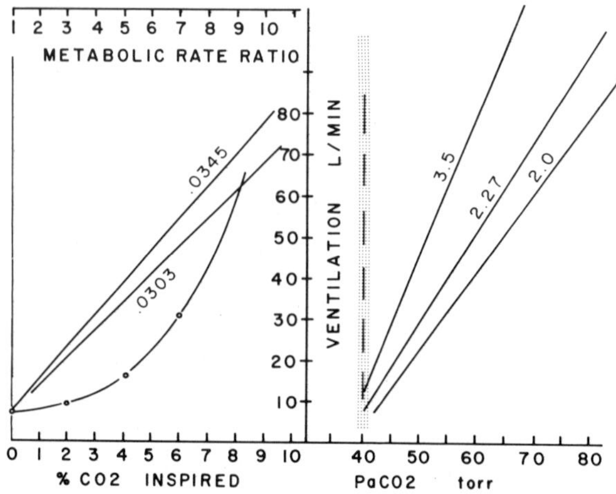

Figure 3.   Performance criteria for the model ventilation in inhalation must be proportional to $PaCO_2$ with slope between 2.0 and 3.5, and isocapnic in exercise[6].

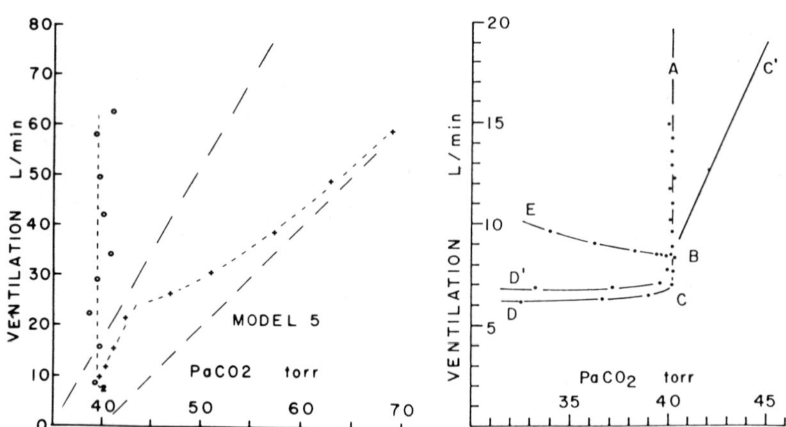

Figure 4 (left).   Model solution using humoral signal only in exercise and inhalation. Note kink in CO_2 response curve[6].

Figure 5 (right).   Model solution using both humoral and neural signals. Note change in scale of ventilation. D, D' are hypometabolic, hypocapnic responses, E is an isometabolic hyperbola[7].

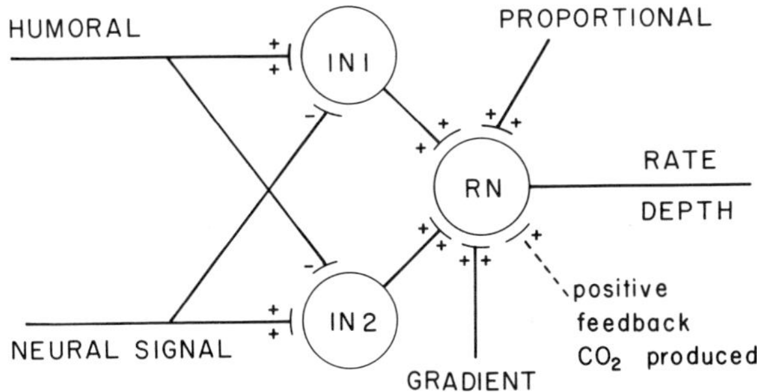

Figure 6. Schematic of controller signal connection to respiratory neurons. Humoral and neural signals in occlusion. Proportional
and gradient signals shown acting on RN are computed as
components of the humoral signal[7].

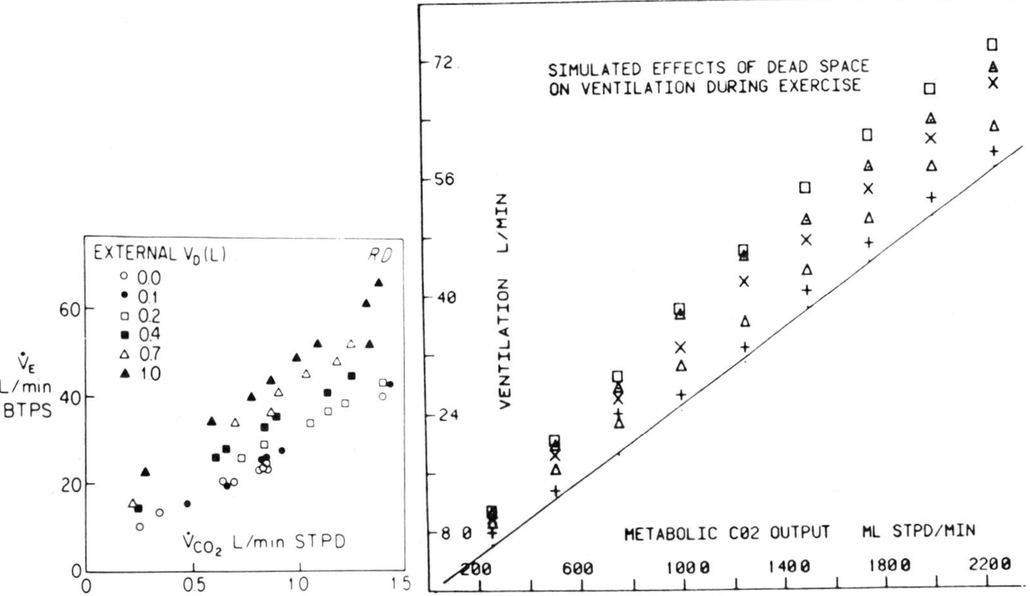

Figure 7. Left. Ventilatory response in exercise with several different
external dead spaces. Magnitudes shown inset Ward and
Whipp[11]. Right. Model simulation using model of Fig.4 and
dead spaces 0, .15, .4, .6, 1.0 and −.05[4].

response and inhalation response are satisfied by any of these assumptions. Since CBF is a function of arterial PCO_2 and it, in turn, is a function of ventilation, the model suggests that CBF can be predicted from the ventilatory response to inhaled CO_2. When the model is run with fixed CBF, an aggregated proportional sensitivity constant in the Cunningham–Lloyd sense can be separated out from blood flow effects. For each fixed sensitivity and blood flow, all CO_2 responses lie on fans of lines with common points at a physically unrealizable negative ventilation[12]. (Fig. 8)

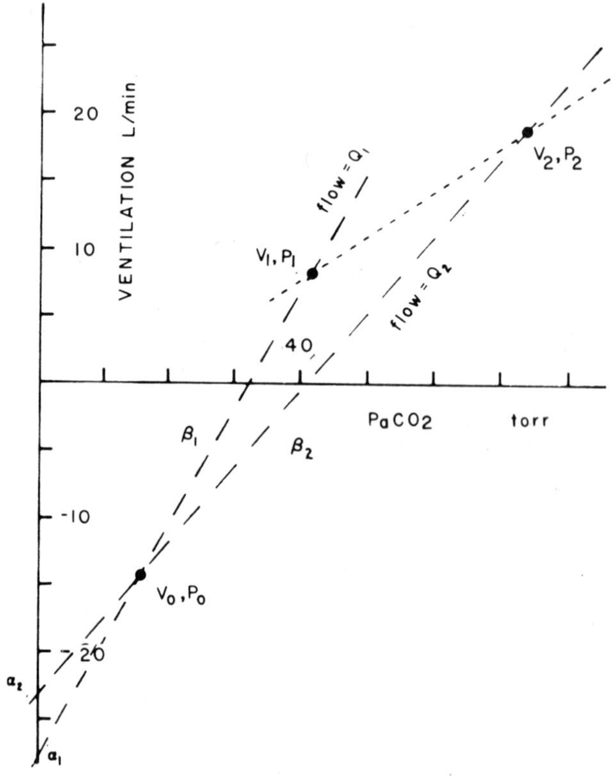

Figure 8.   Nomogram derived from model outputs when forced by CO_2 inhalation and constant blood flows. Sensitivity reduced statistically to single constant S computed from α and β. Blood flow at any point, e.g., V_2, P_2 also a function of α, β[12].

One can construct a nomogram and apply it to compare individuals. Changes in inhaled CO_2 response with age, one deduces, is a decline both in central sensitivity and cerebral blood flow with age (Fig.9). This use validates a model, then, by providing a technological method for study in the laboratory.

ACCEPTANCE

In the end, failure of the community to accept a model's solution to a problem usually rests upon the community's impatience with complexity. Even when they promote parsimony, models become too complicated to explain. What is needed is a computer program to explain other programs. A beginning in this direction is to conceptualize the model as a relational data base system in which the source code is the set of functional dependencies and all named variables are entities in a large table of relations. The source code then generates sets of instantiations of the relations which is the output simulation. Contrary to the rule in the standard conceptualization of a relational data base, there is "row" order between relations in the source and in the output. Interesting new problems arise which pertain to epistemological issues like levels of discourse and obligatory sequences in logic[13,14,15]. As deduced knowledge, physiology is the design of a very large data base, and each laboratory "fact" is a view on instantiated data. The model (physiology) thus reconstructs experimental data, but public conviction seems to require a tutor, an appropriate query language.

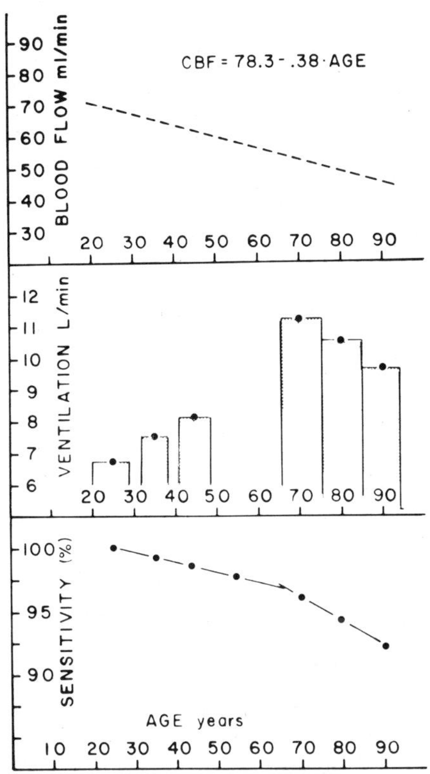

Figure 9. Computed estimates for central CO_2 sensitivity and cerebral blood flow from CO_2 inhalation data on persons of different age[12]. Demonstrates use of model in interpreting data.

Thus, we bring to close the "How and Why", with the Why identified as the entire problem of physiological explanation. I am reminded of a visit to a darkened room in a large museum where a whole, large wall is inscribed entirely with symbols in an ancient language (Fig.10). With a flashlight one can only illuminate a small view (an experiment), one believes that all on the wall deals with one subject (the discipline), and yet one knows that there was a logic that drove the scribe who made the wall (the data base logic). In the end, we need

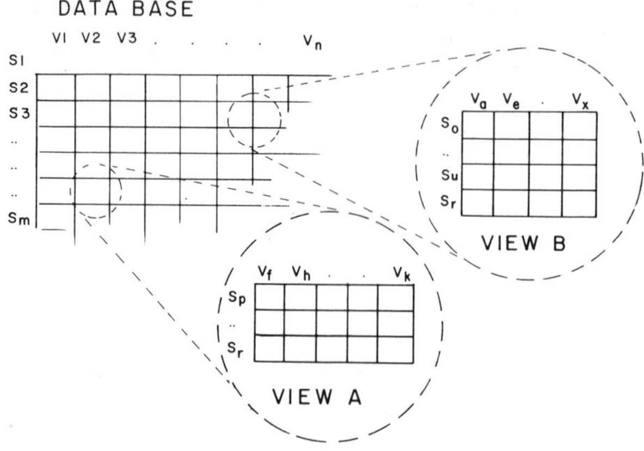

Figure 10. Diagram of either the source code or its computed output as a relational data base instantiated on the wall. The V's are variable names, the S's are either relations or instances of relations. Projected circles are views on the data[15].

something to tell us why we believe the explanations for phenomena. Many modern scholars seem in too–great haste to contemplate complicated explanations and prefer to play in the laboratory as if knowledge itself is power.

REFERENCES

1. B. Benchetrit, P. Baconnier and J. Demongeot, eds., "Concepts and Formalization in the Control of Breathing," Manchester University Press, (1987).

2. B. Whipp and D.M. Wiberg, eds., "Modelling and Control of Breathing," Elsevier Science Pub., New York (1983).

3. W. S. Yamamoto and W. F. Raub, Models of the regulation of external respiration in mammals, Comput. and Biomed. Res. 1:65–104 (1967).

4. W. S. Yamamoto, Information systems approach to integrated responses in the respiratory control system, Ann. Biomed. Eng. 11:349–360 (1983).

5. W. S. Yamamoto and E. S. Walton, On the evolution of the physiological model, Ann. Rev. Biophysics & Bioengin. 4:81–102 (1975).

6. W. S. Yamamoto, A mathematical simulation of the hyperpneas of metabolic CO_2 production and inhalation, Am. J. Physiol. 235:R265–R278 (1978).

7. W. S. Yamamoto, Computer simulation of ventilatory control by both neural and humoral CO_2 signals, Am. J. Physiol. 238:R28–R35 (1980).

8. E. A. Phillipson, J. Duffin and J. D. Cooper, Critical dependence of respiratory rhythmicity on metabolic CO_2 load, J. Appl. Physiol. 50:45–54 (1981).

9. W. S. Yamamoto, Model of steady state breathing at rest and responses to low CO_2 levels, Fed. Proc. 38:3 (1979).

10. M. I. Cohen, Neurogenesis of respiratory rhythm in the mammal, Physiol. Rev. 59:1105–1173 (1979).

11. S. A. Ward and B. J. Whipp, Ventilatory control during exercise with increased external dead space, J. Appl. Physiol. 48:225–231 (1980).

12. W. S. Yamamoto and W. D. Kirk, Model analysis of steady–state ventilatory response to CO_2 into component factors, J. Appl. Physiol. 60:2128–2134 (1986).

13. W. S. Yamamoto and P. G. Wolff, On the identification of verbs in computer programs of physiological models, Comput. & Biomed. Res. 17:175–184 (1984).

14. W. S. Yamamoto, Converting mathematical models of physiological systems to relational data base schemes for analysis and comparison, IEEE Trans. on Biomed. Eng. 32:273–276 (1988).

15. W. S. Yamamoto, Analogies between models of physiological systems and relational data bases, Proc 6th Annual Symposium on Computer Applications in Medical Care pp. 879–881 (1982).

Figures 1 through 9 are reproduced with permission as cited below:
Figure 1 from Ref. 3 in Computers and Biomedical Research.
Figures 2, 4, 5, 6 from Ref. 6 and 7 in the American Journal of Physiology.
Figures 3, 7, 8, 9 from Refs. 8, 11, 12 in the Journal of Applied Physiology.
Figure 7, right half, from Ref. 4 in the Annals of Biomedical Engineering.

ANALYSIS OF THE EXERCISE HYPERPNEA USING DYNAMIC WORK RATE FORCINGS

Richard Casaburi

Division of Respiratory and Critical Care Physiology and Medicine,
Harbor–UCLA Medical Center, Torrance, CA 90509

Though computer modeling of physiological processes has yielded substantial insights, it can be argued that the most important contribution of mathematical modelling is to suggest appropriate laboratory experiments. This was asserted some years ago by Ralph Kellogg[1] in his reference to the seminal work of John Gray[2]. Though the terminology is dated, the message is clear.

> "Perhaps the greatest benefit derived from Gray's writing is that they present his conclusions in such clear, economical, quantitative terms, ... that they have attracted widespread attention and challenged other investigators to devise critical experiments to confirm or refute them".

It is painfully apparent that we do not have adequate information to complete a model of ventilatory control during exercise. Though we know more about how ventilation responds than we did even a few years ago, we are still stuck with the same old question: what are the stimulus–response relationships underlying the exercise hyperpnea? We know that, during the steady–state of moderate exercise in man, arterial P_{CO2}, pH and P_{O2} are not appreciably different from resting levels[3,4]. It would therefore seem reasonable to suppose that keeping the blood gas composition of the arterial blood constant is the primary dictum underlying ventilatory control during exercise. Exactly how this is accomplished is not at all clear.

One of the main strategies which has proven helpful in examining hypotheses has been observing the dynamic characteristics of exercise response. The time course of ventilatory response to a variety of work rate forcings have been examined: step[5,6,7], ramp[8], pulse[9], sine[9–12] and pseudo–random binary[13]. Moreover it has been helpful to compare the time course of ventilation with the time courses of other physiologic variables, oxygen uptake and carbon dioxide output in particular[10,11].

The response to abrupt change in work rate has been more extensively studied than other work rate forcings and it has been helpful to divide the ventilatory response to this stimulus into three phases (Figure 1). The hope is that, if the mechanism of response in each of these three phases can be understood, the overall regulatory scheme will be clear. Phase I is a comparatively rapid increase in ventilation which commences immediately after exercise begins. Phase II is an approximately exponential phase which becomes apparent about 20 seconds after exercise begins. For moderate work rates, ventilation reaches a steady–state within 3–4 minutes. For heavier exercise levels, a superimposed component (Phase III) is observed which appears either as a slower exponential phase or as a progressive "drift" which continues until the subject is exhausted.

<u>Phase I</u>

The Phase I ventilatory response is by far the most difficult to reliably observe, in part because it is so easily volitionally controlled. It is most prominent in the rest–to–exercise transition, but it has been necessary to resort to averaging on the order of 50 responses to identical stimuli to get good dynamic resolution[14,15]. The reason why such stress has been placed on such a subtle component can be directly traced to the 1913 work of Krogh and Lindhard[6] . Observing that ventilation increased with no measurable latency after exercise started, they had the insight to conclude:

> "If the excess of CO_2 produced in the muscles (at exercise onset) were responsible for the rise in ventilation, there must be a latent period until the blood could reach the respiration centre... The mechanism which shall produce the abrupt changes must be a nervous mechanism".

This interpretation is, in essence, the foundation of the neural theories of the exercise hyperpnea[17]. These theories hold that neurally mediated signals either from the exercising limbs[18,19], or from higher brain centers[16,20], impinge directly on the respiratory center in the medulla.

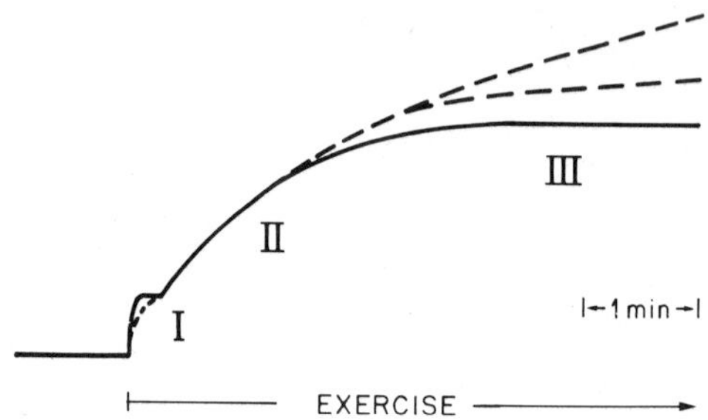

Fig. 1.　Schematic representation of the three phases of response to the onset of exercise.

Over the years, a large number of studies have been performed, both in animal and in man, to evaluate the contribution of neurally mediated signals from either central or peripheral sources[1]. Though this remains an area of controversy, no unequivocal evidence has been obtained that either source plays an appreciable role in the Phase I hyperpnea[22,23].

In recent years, additional hypotheses have been advanced to explain a ventilatory response before the exercising leg to lung circulatory delay has elapsed. One of the more interesting hypotheses has been that the abrupt ventilatory increase is in some way linked to the rapid cardiac output increase that is seen as exercise starts (although the receptors which could mediate this linkage are not apparent). Our laboratory, and some others, have reported situations in which experimental manipulation of the initial cardiac output increase has been accompanied by similar changes in ventilation[14,25]. However, we recently reported a study which was different in that the initial cardiac output change was altered without giving perceptual cues to the subject[15].

We studied patients who had previously had cardiac pacemakers inserted for treatment of complete heart block. In these otherwise healthy patients, heart rate was completely controlled by the pacemaker. An external pacemaker controller was used to reset heart rate in an unobtrusive manner. The strategy was to set the heart rate low (50 beats/min) while the subject sat at rest on the cycle ergometer with the pacemaker programmer held over the precordium. Under these circumstances, stroke volume would be forced to be relatively high. As exercise started, heart rate was either held constant at 50 beats/min or abruptly increased to 100 beat/min using the pacemaker programmer. In those transitions from rest to exercise in which heart rate was held constant, there should be limited initial cardiac output response, because of limited increase in stroke volume. In those transitions in which heart rate was accelerated, cardiac output should be accelerated. Figure 2 shows that this strategy was successful. In this figure, the responses of each of 5 subjects who performed 8 of each of the two kinds of transitions were averaged (i.e., each curve is the average of 40 responses). The left panel presents the oxygen uptake response in the first 4 minutes after exercise onset. The Fick relation tells us that, at a given arteriovenous O_2 content difference, oxygen uptake will be in direct proportion to cardiac output. Thus in the initial 30 seconds of exercise, the fact that oxygen uptake was distinctly lower when heart rate was held constant (dashed line) than when heart rate was accelerated (solid line) means that the initial cardiac output response must have been cut roughly in half by holding heart rate constant. The right hand panel of Figure 2 shows that, despite substantial differences in cardiac output time course, Phase I ventilation is not discernibly different. These results seem inconsistent with a strong -link between cardiac output and ventilation.

This, then, is a very confusing situation, with evidence against all of the proposed physiologic mediators of the Phase I exercise hyperpnea. A study recently reported from this laboratory makes the situation even more murky. As explained above, the central assumption about the Phase I ventilatory response is that it occurs before blood bourne signals caused by the exercise reach the central circulation. Interestingly, this assumption has never been explicitly validated and we set out to provide this validation. We utilized a technique which allowed catheterization of the pulmonary artery of normal subjects via an antecubital approach. These subjects then performed transitions from rest to moderate exercise on a

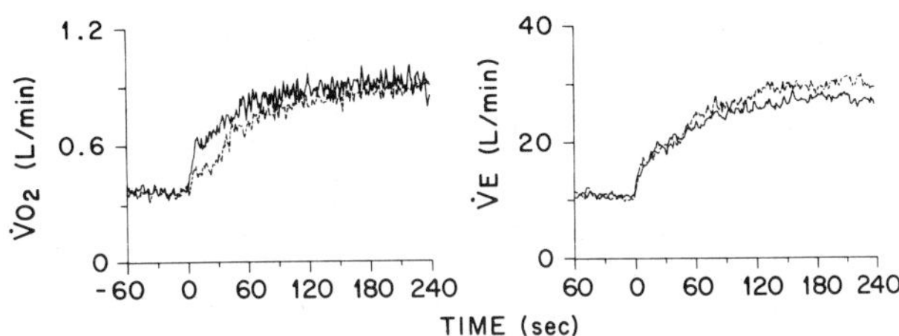

Fig. 2.   Average response of 5 subjects with implanted cardiac pacemakers to the onset of exercise (each curve is the second–by–second average of 40 responses). Transition from rest to exercise occurred at time "0". Solid lines show responses to studies in which pacemaker rate was abruptly increased at time "0" from 50 to 100 beats per minute. Dashed lines are studies in which pacemaker rate was held constant at 50 beats per minute. Left panel: oxygen uptake response. Right panel: ventilatory response. (see text) (from Ref. 15).

cycle ergometer. A computerized anaerobic sampling manifold was used to withdraw blood
from the pulmonary artery catheter. During each transition, the manifold drew 19 blood
samples, usually at a rate of one sample every 4 seconds. These blood samples underwent
blood gas analysis so that a high resolution time course of mixed venous blood gas changes
that occur as exercise starts was obtained. Figure 3 presents the time course of oxygen saturation and P_{CO_2} of the mixed venous blood from 20 seconds before to 45 seconds after the
onset of moderate exercise (the data presented represents the average of 12 transitions in 6
subjects). What we had expected to see was constancy of these variables for perhaps 15–20
seconds, until the leg–to–lung circulation time had elapsed[27]. Contrary to these expectations,
Figure 3 shows that clear changes take place in both O_2 saturation and P_{CO_2} within a few
seconds of exercise onset. The mechanism of these changes is unclear as is the potential for
these humoral changes to play a role in the early ventilatory response to exercise. Yet it is
apparent that the assumption that no humoral changes reach the central circulation during
Phase I is not warranted. Very recent studies have yielded the interesting finding that these
abrupt changes in mixed venous blood gas composition seen in the transition between rest
and exercise are not observed in other modes of exercise transition[28].

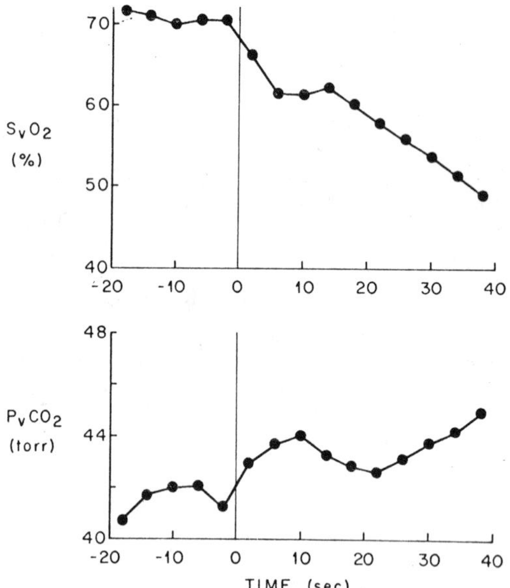

Fig. 3.   Averaged time course of pulmonary arterial blood com-
position at the transition from rest to exercise (N = 12).
Top: mixed venous O_2 saturation. Bottom: mixed venous
P_{CO_2}. Mixed venous O_2 saturation decreases and P_{CO_2} in-
creases with no apparent latency after exercise onset (from
Ref. 26).

Phase II

The Phase II ventilatory response can be observed more or less in isolation by observing
the response to transitions between work rates, for example, the transition between unloaded
pedalling and exercise or by observing the response to sinusoidal variation of work rate. These
responses have very little Phase I component. If work rates are restricted to the moderate
range, Phase III responses can be avoided.

Some 10 years ago, we published studies of the responses to sinusoidal work rate variation, using Fourier analysis techniques to determine the kinetics of the ventilatory and gas

exchange responses[10,11]. Figure 4 shows that, for a group of 10 normal subjects, the time constant of ventilation is closely correlated with but longer than the time constant of CO_2 output, indicating that ventilatory changes lag CO_2 output changes slightly. This finding is reinforced if the time course of arterial P_{CO2} in response to sinusoidal work rate is examined. Arterial P_{CO2} shows a small but significant fluctuation with work rate and, importantly, P_{aCO2} is high when ventilation is high[29]. This suggests that Phase II ventilation is chemoreceptor linked – specifically through CO_2 sensitivity.

Several experimental interventions have been found to slow Phase II ventilatory kinetics. For example, hyperoxia slows the response to sinusoidal exercise: the time constant of ventilatory response becomes longer[10] . Similarly, dopamine infusion[30], induced metabolic alkalosis[31] and carotid body resection[32] all result in slow ventilatory kinetics. What these interventions have in common is that the carotid body contribution to respiratory chemoreception is reduced. The carotid bodies apparently contribute to the tight control of arterial P_{CO2} during Phase II.

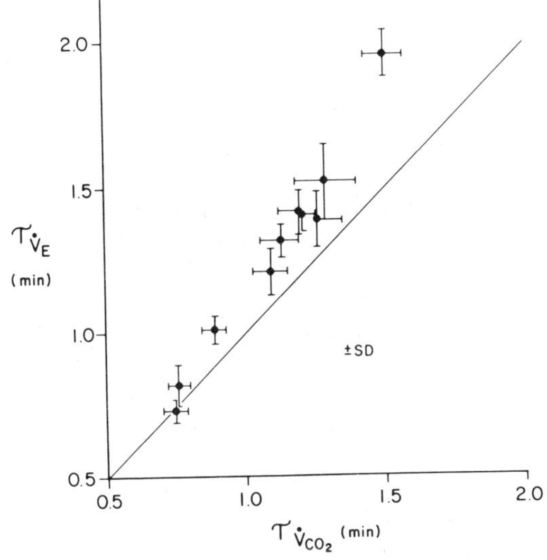

Fig. 4.   Correlation of ventilatory and carbon dioxide output dynamics. Time constants of response of 10 normal subjects were determined from the responses to sinusoidal work rate variation. (from Ref. 14).

An important issue in the mathematical modelling of ventilatory response is whether the conditions of linearity are satisfied. In particular, it is important to know whether the dynamics of ventilatory response are strongly influenced by work rate. If so, the comparisons of ventilatory kinetics among subjects[32,33] and between species[34] become much more difficult to interpret. A priori, it might be suspected that the complex interaction of intramuscular metabolic processes, humoral transport delays and chemoreceptor responses would mitigate against dynamic linearity. To address this issue, normal subjects performed a large series of transitions from unloaded pedalling to one of 7 work rates spaced evenly below the maximum the subject was able to tolerate[7]. Figure 5 shows the ventilatory responses of one subject; each curve represents the average of breath–by–breath responses to 5 to 10 exercise transitions. It can be seen that a steady–state in ventilation is reached by 3 or 4 minutes at the lowest few work rates, but the steady–state is delayed at the higher work rates – a clear evidence of a non–linear response. To quantitate the ventilatory response kinetics, a single exponential was fit to each response by iterative non–linear least squares regression techniques. Figure 6 shows the variation of the time constant of ventilatory response for the 7 progressively increasing work rates in 4 normal subjects. The ventilatory time constant is invariant at about 75 seconds over the lower 4 work rates, but is much longer at the 3 higher work rates. The upper panel presents the average end–exercise blood lactate at each of the 7 work rates. It can be seen that at work rates engendering lactic acidosis, ventilatory kinetics are progressively slowed. Over the range of work rates not associated with lactic acidosis, the ventilatory response appears to exhibit dynamic linearity. It would seem that evaluation of ventilatory kinetics requires knowledge of the blood lactate level associated with the exercise.

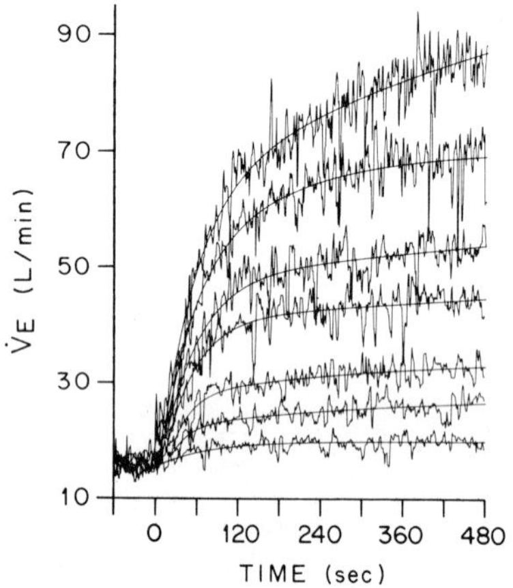

Fig. 5. Second–by–second ventilatory responses of a normal subject to transitions between unloaded pedalling and each of 7 higher work rates evenly spaced below 205 watts. Each curve is the average of 4–8 transitions. Superimposed on each curve is the best fitting model relation, as determined by non–linear regression techniques (from Ref. 7).

Phase III

Understanding the Phase III ventilatory response requires definition of the stimulus causing the slow upward drift in ventilation which occurs at high work rates. Several known ventilatory stimuli come into play at high work rates: elevated blood lactate[35] and catecholamine[36] levels and elevated body temperature[37]. To identify the predominant stimulus to the Phase III ventilatory response, we capitalized on the observation that a program of endurance exercise training reduces the ventilatory requirement for heavy exercise[38]. Figure 7 demonstrates this nicely. The ventilatory response of a normal subject to 4 work rates, ranging from moderate to very heavy are shown. After 8 weeks of training, the response to identical work rates features a dramatic decrease in the ventilatory response at the higher exercise intensities. In the 10 subjects we studied, ventilation was reduced by an average of 35 liters/minute at the highest work rate studied, but very little at the lowest work rate. When we examined how the potential ventilatory stimuli had been influenced by training, it was apparent that, at the higher work rates, all of the possible mediators we measured (blood lactate, epinephrine and norepinephrine as well as body temperature) had been reduced by training. However, when the correlations between the training–induced decreases in these possible mediators and the decrease in ventilation were examined, one correlation was very

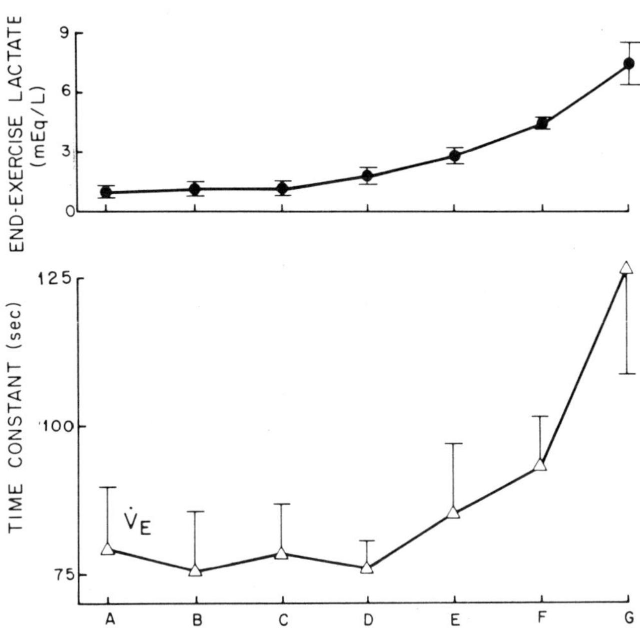

Fig. 6.   Bottom: relationship between the average ($\pm$ 1 S.E) time constant of ventilatory response and exercise intensity. Seven work rates were spaced evenly below exercise intensity G (the highest tolerated) for each of 4 subjects. Top: average ($\pm$ 1 S.E.) end–exercise blood lactate level in the 4 subjects at each exercise intensity. The time constant of $\dot{V}_E$ is markedly longer at work rates associated with lactic acidosis (from Ref. 7).

much stronger than the others. The changes in ventilation and blood lactate were highly correlated (Fig. 8); on average ventilation fell 5.7 liters/min for each meq/L fall in blood lactate. Thus it seems likely that the Phase III ventilatory response is linked to ventilatory stimulation by lactic acidosis. This observation may turn out to have clinical importance. Though normal subjects are not limited in their exercise tolerance by the ventilation they can sustain, patients with obstructive lung disease are. These patients may be able to benefit

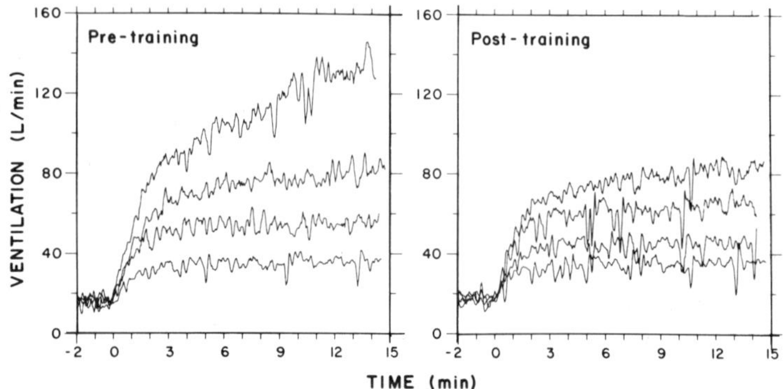

Fig. 7.   Effect of endurance training on time course of ventilation in a normal subject. A, pretraining breath–by– breath responses to work rates of 95, 148, 191 and 233 watts. B, post training responses to identical work rates after 8 weeks of training. Note that training produces a dramatic decrease in ventilation at the higher work rates (from Ref. 38).

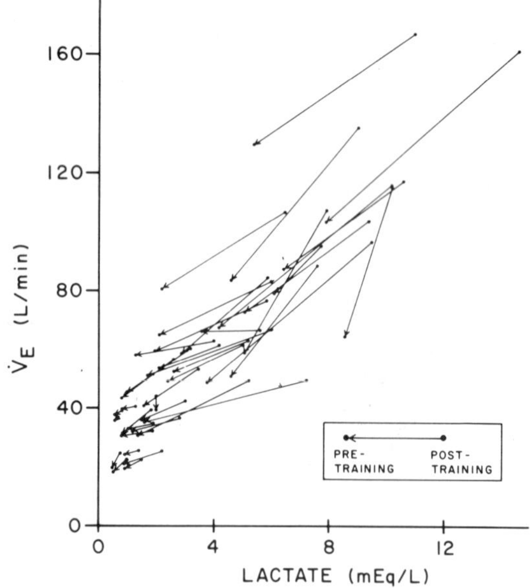

Fig. 8.   Relation between reduction in end–exercise ventilation and reduction in end–exercise lactate produced by endurance training. Lines connect before and after training responses to identical exercise tests (Arrowhead points to after training responses) (from Ref. 38).

20

from a program of exercise training designed to reduce the blood lactate produced by a given level of exercise and thereby reduce the level of ventilation required[39]. Recent results suggest that this may be a workable strategy for increasing exercise tolerance[40].

In summary, a substantial amount is known about the mediation of the kinetics of the ventilatory response to exercise. Phase I is the most enigmatic and may yet be discovered to be a "startle" response, of no great relevance to the regulation of the exercise hyperpnea. Phase II response seems to be CO_2 related; the carotid bodies act to assure tight regulation of arterial P_{CO2}. Phase III of the ventilatory response appears to be caused by the lactic acidosis which accompanies heavy exercise. Despite these insights, it must be acknowledged that the overall regulatory strategy by which the exercise hyperpnea is controlled remains elusive.

REFERENCES

1. R.H. Kellogg, Central chemical regulation of respiration in: "Handbook of Physiology. Respiration Vol. I," W.O. Fenn and H. Rahn, ed, Am. Physiol. Soc., Washington,D.C. (1964).

2. J.S. Gray, "Pulmonary Ventilation and its Physiological Regulation," Thomas, Springfield, IL. 1959, p.82.

3. B.J. Whipp, The control of exercise hyperpnea in: "Regulation of Breathing," T.F. Hornbein, ed., Dekker, New York (1981).

4. K. Wasserman, S.L. VanKessel and G.G. Burton, Interaction of physiological mechanisms during exercise, J. Appl. Physiol. 22:71–85 (1967).

5. D. Linnarsson, Dynamics of pulmonary gas exchange and heart rate changes at start and end of exercise, Acta Physiol. Scand. Suppl. 415:1–68, (1974).

6. G. Mattell. Time–courses of changes in ventilation and arterial gas tensions in man induced by moderate exercise, Acta Physiol. Scand. Suppl. 206:1–53 (1973).

7. R. Casaburi, T.J. Barstow, T. Robinson and K. Wasserman, Influence of work rate on ventilation and gas exchange kinetics, J. Appl. Physiol. 67:547–555 (1989).

8. Y. Fujihara, J.R. Hildebrandt and J.Hildebrandt, Cardiorespiratory transients in exercising man. I. Tests of superposition, J. Appl. Physiol. 35:58–67 (1973).

9. H.K. Bakker, T.R. Struikenkamp, and G.A. DeVries, Dynamics of ventilation, heart rate, and gas exchange: sinusoidal and impulse work loads in man, J. Appl. Physiol. 48:289–301 (1980).

10. R. Casaburi, R.W. Stremel, B.J. Whipp, W.L. Beaver, and K. Wasserman, Alteration by hyperoxia of ventilatory dynamics during sinusoidal work, J. Appl. Physiol. 48:1083–1091, (1980).

11. R. Casaburi, B.J. Whipp, K. Wasserman, W.L. Beaver, and S.N. Koyal, Ventilatory and gas exchange dynamics in response to sinusoidal work, J. Appl. Physiol. 42:300–311 (1977).

12. O. Wigertz, Dynamics of ventilation and heart rate in response to sinusoidal work load in man, J. Appl. Physiol. 29:208–218 (1970).

13. F.M. Bennett, P. Reischl, F.S. Grodins, S.M. Yamashiro, and W.E. Fordyce, Dynamics of ventilatory response to exercise in humans, J. Appl. Physiol. 51:194–203 (1981).

14. R. Casaburi, B.J. Whipp, K. Wasserman, and R. W. Stremel, Ventilatory control characteristics of the exercise hyperpnea as discerned from dynamic forcing techniques, Chest. 73:280S–283S (1978).

15. R. Casaburi, S. Spitzer, R. Haskell and K. Wasserman, Effect of altering heart rate on oxygen uptake at exercise onset, Chest. 95:6–12 (1989).

16. A. Krogh and J. Lindhard, The regulation of respiration and circulation during the initial stages of muscular work, J. Physiol. (London), 47:112–136 (1913).

17. P. Dejours, Control of respiration muscular exercise, in: "Handbook of Physiology. Respiration" Vol I, W.O. Fenn, and H. Rahn eds., Am. Physiol. Soc. Washington, D.C. (1964).

18. T.R. Harrison, W.G. Harrison, Jr., J.A. Calhoun and J.P. Marsh, Congestive heart failure XVII. The mechanism of dyspnea on exertion, Arch. Intern. Med. 50:690–720 (1932).

19. F.F. Kao, An experimental study of the pathways involved in exercise hyperpnea employing cross–circulation techniques, in: "The Regulation of Human Respiration," D.J.C. Cunningham and B.J. Lloyd, ed, Blackwell, Oxford (1963).

20. F.L. Eldridge, D.E. Millhorn and T.G. Waldrop, Exercise hyperpnea and locomotion parallel activation from the hypothalamus, Science 211:844–846 (1981).

21. K. Wasserman, B.J. Whipp, R. Casaburi, Respiratory control during exercise; in: "Handbook of Physiology. The Respiratory System Control of Breathing," A.P. Fishman, ed., Am. Physiol. Soc., Bethesda, (1986).

22. A.G. Brice, H.V. Forster, L.G. Pan, A. Funahashi, T.F. Lowry, C.L. Murphy and M.D. Hoffman, Ventilatory and P_{aCO2} response to voluntary and electrically induced leg exercise, J. Appl. Physiol. 64:218–225 (1988).

23. A.G. Brice, H.V. Forster, L.G. Pan, A. Funahashi, M.D. Hoffman, C.L. Murphy and T.F. Lowry, Is the hyperpnea of muscular contractions critically dependent on spinal afferents?, J. Appl. Physiol. 64:226–233 (1988).

24. K. Wasserman, B.J. Whipp and J. Castagna, Cardiodynamic hyperpnea: hyperpnea secondary to cardiac output increase, J. Appl. Physiol. 36:457–464 (1974).

25. D. Weiler–Ravell, D.M. Cooper, B.J. Whipp and K. Wasserman, Control of breathing at the start of exercise as influenced by posture, J. Appl. Physiol. 55:1460–1466 (1983).

26. R. Casaburi, J. Daly, J.E. Hansen and R.M. Effros, Abrupt changes in mixed venous blood gas composition after the onset of exercise, J. Appl. Physiol. 67:1106–1112 (1989).

27. T.J. Barstow and P.A. Mole, Simulation of pulmonary O_2 uptake during exercise transients in humans, J. Appl. Physiol. 63:2253–2261 (1987).

28. R. Casaburi, C. Cooper, R.M Effros and K. Wasserman, Time course of mixed venous

oxygen saturation following various modes of exercise transition, FASEB J. 3:849 (1989).

29. B. J. Whipp, K. Wasserman, R. Casaburi, C. Juratsch, M.L. Weissman and R.W. Stremel, Ventilatory control characteristics of conditions resulting in isocapnic hyperpnea, in: "The Regulation of Respiration During Sleep and Anesthesia" R.S. Fitzgerald, H. Gautier, and S. Lahiri, ed, Plenum, New York, (1978).

30. C.L. Boetger and D.S. Ward, Effect of dopamine on transient ventilatory response to exercise, J. Appl. Physiol. 61:2102–2107 (1986).

31. A. Oren, B.J. Whipp, K. Wasserman, Effect of acid–base status on the kinetics of the ventilatory response to moderate exercise, J. Appl. Physiol. 52:1013–1017 (1982).

32. K. Wasserman, B.J. Whipp, S.N. Koyal, M.G. Cleary, Effect of carotid body resection on ventilatory and acid–base control during exercise, J. Appl. Physiol. 39:354–358 (1975).

33. L.B. Diamond, R. Casaburi, K. Wasserman and B.J. Whipp, Kinetics of gas exchange and ventilation in transitions from rest or prior exercise, J. Appl. Physiol. 43:704–708 (1977).

34. R. Casaburi, M.L. Weissman, D.J. Huntsman, B.J. Whipp and K. Wasserman. Determinants of gas exchange kinetics during exercise in the dog, J. Appl. Physiol. 46:1054–1060 (1979).

35. S.N. Koyal, B.J. Whipp, D. Huntsman, G.A. Bray and K. Wasserman. Ventilatory responses to the metabolic acidosis of treadmill and cycle ergometry. J. Appl. Physiol. 40:864–867 (1976).

36. R.J.. Butland, J.A.C.K. Pang and D.M. Geddes, Effect of B–adrenergic blockade on hyperventilation and exercise tolerance in emphysema, J. Appl. Physiol. 54:1368–1373 (1983).

37. J.M. Hagberg, J.P. Mullin, and F.J. Nagle, Oxygen consumption during constant–load exercise, J. Appl. Physiol. 45:381–384 (1978).

38. R. Casaburi, T.W. Storer and K. Wasserman, Mediation of reduced ventilatory response to exercise after endurance training, J. Appl. Physiol. 63:1533–1538 (1987).

39. R. Casaburi, K. Wasserman, A. Patessio, F. Ioli, S. Zanaboni and C.F. Donner, A new perspective in pulmonary rehabilitation:anaerobic threshold as a discriminant in training, Eur. Respir. J. 2 Suppl 7:618s–623s (1989).

40. R. Casaburi, A. Patessio, F. Ioli, S. Zanaboni, C. Donner and K. Wasserman, Reductions in exercise lactic acidosis and ventilation after exercise training in obstructive lung disease patients, Am. Rev. Respir. Dis. 139:A330 (1989).

OPTIMAL REGULATION OF VENTILATION DURING EXERCISE

C.-S. Poon

Harvard–M.I.T. Division of Health Sciences and Technology
Cambridge, Massachusetts 02139

INTRODUCTION

Perhaps one of the most long–standing tenets in physiological research is the general belief that every biological response must stem from a specific stimulus. Beginning with Sherrington[1], this "stimulus–response" or reflex control paradigm has continued to dominate current models of regulatory processes in the body. In the long history of respiratory physiology such a view has prevailed. For example, in the well–celebrated Hering–Breuer reflex one finds a direct association of the stimulus (pulmonary stretch receptor activity) with the resulting response (delayed onset of inspiration). Similarly, in the chemoreflex regulation of respiration one observes an increase in ventilation in response to chemoreceptor stimulation. In both these cases the picture seems quite clear: there is a unique, one–to–one relationship between the purported stimulus and the resulting response. Thus, a causal effect appears to be evident, and the simple reflex scheme as a general physiological control paradigm appears to work remarkably well. There is little reason to warrant any alternative explanation for a simple empirical relationship that seems so trivial and obvious.

That may be so. But there are circumstances in which the reflex control scheme is known to fall short. During muscular exercise respiratory ventilation typically rises in direct proportion to metabolic rate with little change in any of the known chemical stimuli. This condition persists in the steady state over a wide range of metabolic rate. Thus, granting the validity of the chemoreflex model in describing the effects of chemical stimulation, one is faced with the dilemma of having to account for the exercise hyperpnea which obtains in the face of an apparent chemical homeostasis in the arterial blood. On the other hand, if one accepts the etiological view that the chemical feedback mechanism is designed simply to maintain arterial chemical homeostasis (e.g. via a constant CO_2 set point), then one must explain the observed disruption of arterial blood gas and pH balance during CO_2 inhalation. Either way, the simple chemoreflex model seems to be less than satisfactory.

To resolve this dilemma there are at least two alternative theoretical approaches. The first one, which is the simplest and most straightforward, is to continually acknowledge the structure of the chemoreflex model but hypothesize the existence of an additional, super-imposed "exercise stimulus" that reflexly sustains an increase in ventilation during exercise. This model of separate feedback and feedforward control for chemoreflex and exercise re-flex, respectively, was first explicitly formulated by Grodins[2] and perhaps tacitly assumed by most experimental investigators. Although conceptually satisfying, this approach lacks the experimental support of a realistic physiological mechanism that can be identified with the hypothetical feedforward pathway. Despite extensive search by numerous investigators for over a century, the putative exercise stimulus so far has remained elusive. So there is a major void in the theory that is yet to be filled, and the search for the exercise stimulus goes on.

The assumption of a reflex mechanism either by feedback or feedforward control is tantamount to the assertion that the brainstem respiratory center amounts to – in control terminology – nothing more than a simple proportional controller (Fig. 1a). That is, its input–output relationship is fixed and is given by a simple ratio. Such a controller is said to be a "dumb" controller because it only passively follows the input. The required operation represents a relatively low level of neural integration and information processing in comparison with the complex neuronal organization of the brain stem.

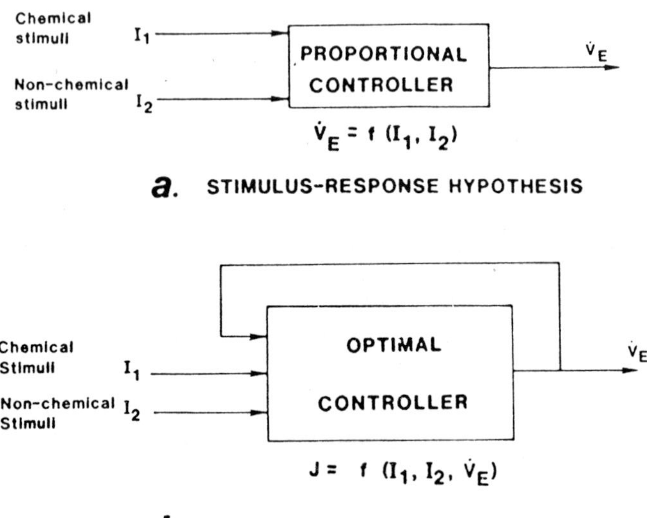

a. STIMULUS-RESPONSE HYPOTHESIS

b. OPTIMAL CONTROL HYPOTHESIS

Fig. 1. Hypothetical structures of respiratory controller. (a) reflex control; (b) optimal control. (From Poon[3]).

The second theoretical solution to the exercise hyperpnea dilemma is not as straight-forward. It requires a breaking down of conventional beliefs and rebuilding of the model from square one. Rather than looking for an unknown exercise stimulus, the question being asked is: Could the CO_2 and exercise responses be engendered by a different central mechanism than a simple reflex controller? It has long been recognized that under some circumstances breathing pattern may be regulated by the controller in such a way that the total work rate of breathing is minimized[4]. In other words, there is an optimization policy with respect to the work output of the respiratory neuromuscular system. This represents a different kind of control paradigm – one that is more sophisticated and probably requires a higher level of neural integration.

Adapting the optimization paradigm to the control of ventilation requires a leap of faith. So far, the concept of optimization has been restricted to the regulation of breathing pattern, with ventilation always being assumed to be constant – as though it were something sacred, a constraint that is set only by the chemical and metabolic demands of the body. But why? If breathing pattern could be subject to optimization, why not ventilation?

Short of any excuses, I found it necessary that we give the ventilatory optimization model a fair trial by first building one. Figure 1b shows the structure of such a hypothetical controller for the optimization of ventilation. The controller receives both chemical and mechanical feedbacks by way of the conventional chemosensory and mechanosensory pathways. The control strategy is to regulate ventilatory output such that arterial chemical homeostasis is maintained as much as possible without incurring excessive respiratory work output. Thus, the controller is not required to follow a unilateral command (as does a reflex controller) but may respond to conflicting demands by balancing an overall objective. As such it is a "smart" or intelligent controller because it has the ability to adapt, evaluate, and optimize.

In the following, I will present the detailed structure and principle of operation of the optimization model and demonstrate its validity in describing the CO_2 and exercise ventilatory responses. It is shown that the model exhibits some interesting characteristics which distinguish it from the conventional models. Also summarized are some recent experimental findings which help corroborate the optimization model.

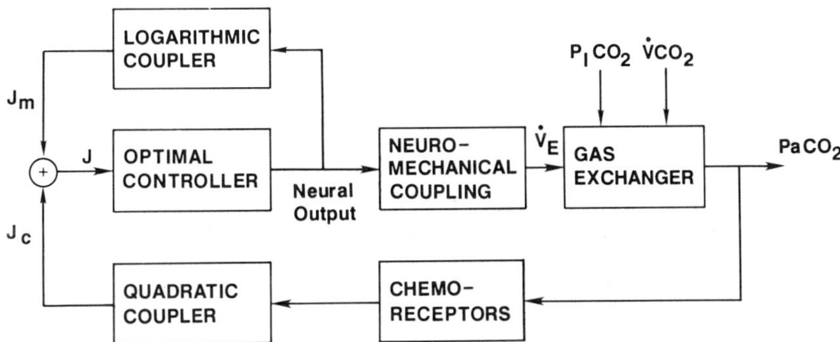

Fig. 2. Complete respiratory optimization model. (From Poon[5]).

MODEL DESCRIPTION

As in any closed–loop control system, the complete model is composed of three parts: controller, controlled system and feedback system (Fig. 2). The controlled system is the pulmonary gas exchanger in which arterial blood gas tensions and pH are directly controlled by $\dot{V}_E$. For simplicity of discussion only the CO_2 component is considered here with the assumption that any hypoxic and acidotic effects are either negligible or being held constant. For a uniform lung compartment with no shunt the steady–state input–output relationship is given by the well–known CO_2 metabolic equation:

$$P_a CO_2 = P_I CO_2 + \frac{863\ \dot{V}CO_2}{\dot{V}_E(1 - V_D/V_T)} \tag{1}$$

The variables $P_I CO_2$ and $\dot{V}_{CO2}$ are considered to be disturbance signals to the controlled system. Changes in these factors will alter the input–output relationship in different ways.

$$\text{Table 1. Glossary of Symbols}$$

$\dot{V}_E$	Pulmonary ventilation (L/min., BTPS)
$\dot{V}CO_2$	CO_2 output (L/min, STPD)
V_D/V_T	Dead space-to-tidal volume ratio
$\dot{V}_{max}$	Maximum ventilation (L/min, BTPS)
$\dot{V}_{Eo}$	Ideal ventilatory response (L/min, BTPS)
$PaCO_2$	Arterial CO_2 tension (Torr)
P_ICO_2	Inspired CO_2 partial pressure (Torr)
P_{100}	Inspiratory mouth occlusion pressure at 100 ms
J_c	Perceived chemical cost of controller
J_m	Perceived mechanical cost of controller
J	Total control cost
α	Apparent CO_2 sensitivity of chemoreceptors
β	Apparent CO_2 threshold of chemoreceptors
IRL	Inspiratory resistive load
IEL	Inspiratory elastic load

The feedback system consists of two separate pathways. The chemosensory pathway detects changes in $PaCO_2$ by way of the conventional peripheral and central chemoreceptors. The assumptions are that the chemoreceptor activities are linearly related to $PaCO_2$ and the outputs of the peripheral and central chemoreceptors are directly additive. Furthermore, the controller's perception of the total chemosensory feedback is assumed to follow Steven's power law[6] with a square exponent:

$$J_c = \{\alpha(PaCO_2 - \beta)\}^2 \tag{2}$$

The square law is an empirical relationship that is chosen to conform with experimental ventilatory response data.

The mechanosensory pathway detects changes in the neuro–mechanical output of the effector organs and relays it to the controller. The respiratory system is known to be endowed with a rich supply of mechano receptors arising variously in the respiratory muscles, chest wall, lungs and airways, and there may be neuronal mechanisms for the direct sensing of the neural outflow of the respiratory motor units. The existence of a mechanosensory pathway is therefore unequivocal, and its role in modulating breathing pattern is well documented. It has not been clear, however, whether such a mechanical pathway is directly involved in ventilatory control. The present model postulates that mechanosensory feedback is an integral part in the optimization of $\dot{V}_E$. Specifically, it assumes that the mechanical cost of breathing perceived by the controller increases logarithmically with $\dot{V}_E$:

$$J_M = 2\log \frac{\dot{V}_E}{(1 - \dot{V}_E/\dot{V}_{max})} \tag{3}$$

The logarithmic relation follows the classical Weber–Feber law of sensory perception[6]. In reality, the perceived intensity is likely to be a complex function of respiratory neuromechanical output. For the purpose of optimization of ventilation, however, it suffices to express the perceived feedback signal as a simple function of $\dot{V}_E$. The factor $(1 - \dot{V}_E/\dot{V}_{max})^2$ in Eq. 3 accounts for the decrease in neuromechanical efficiency as $\dot{V}_E$ increases. It has a maximum value of 1 (100% efficiency) for $\dot{V}_E = 0$ and decreases monotonically to 0 as $\dot{V}_E$ approaches $\dot{V}_{max}$, where the system becomes mechanically limited. The precise variation of the neuromechanical efficiency for intermediate values of $\dot{V}_E$ is not completely clear. Nevertheless, the proposed factor may serve as a first approximation for the purpose of predicting $\dot{V}_E$.

The central hypothesis of the optimization model is that ventilatory output is set by the controller so as to minimize an overall operating cost of the form

$$J = J_c + J_m \tag{4}$$

which is simply an algebraic sum of the chemical and mechanical costs as perceived by the controller. As ventilation rises, more mechanical effort is expended and thus J_m goes up. However, the resultant lowering in $PaCO_2$ which accompanies the hyperventilation will in turn cause J_c to go down, thereby offsetting or even reversing any net increase in J. Depending on the direction of change of J, an increase in $\dot{V}_E$ may or may not be beneficial for cost minimization. An optimal response under any given condition is such that any further increase or decrease in $\dot{V}_E$ with unchanging inputs would eventually lead to an increase in J.

An important point to notice here is that under this operating rule the ventilatory response no longer directly follows the feedback "drives". Rather, it follows the directional gradient of the cost profile until it rests in equilibrium at the trough of the cost–action terrain. In other words, the "drive" or "incentive" for any changes in $\dot{V}_E$ is derived from a potential lowering in J as a result of the change; action is taken only if it will contribute to a lowering in overall cost.

Such a control strategy has the advantage of offering greater adaptability to different challenges than does a reflex controller. For example, during CO_2 breathing the ability of the lung to clear CO_2 is impaired by the influx of inhaled CO_2 which increases with ventilation. In order to curtail the resulting rise in $PaCO_2$ a relatively large increase in $\dot{V}_E$ is required. As an extreme case, it would take an infinitely large ventilation to bring $PaCO_2$ close to the inspired PCO_2 level. Thus the "price" in terms of J_m is high, and the "reward" in terms of a lowering in J_c is quite limited. A prudent response is therefore to allow a moderate elevation in $PaCO_2$ while avoiding an excessive increase in $\dot{V}_E$. The resulting changes in $\dot{V}_E$ and $PaCO_2$ in this case resemble the hypercapnic ventilatory response which is similarly predicted by the reflex model. By contrast, during exercise the rate of elimination of pulmonary CO_2 increases directly with $\dot{V}_E$, and any potential increase in $PaCO_2$ can be contained relatively effectively by augmenting $\dot{V}_E$. To cope with the elevated metabolic rate the controller has two options. The first is to stay put and maintain the same $\dot{V}_E$, in which case $PaCO_2$ would rise sharply thereby incurring an excessively large J_c. The other option is to increase $\dot{V}_E$ while keeping $PaCO_2$ constant, thereby incurring an increase in J_m. On balance, the latter option turns out to be more favorable because it is cost effective under this circumstance to expend some mechanical work so as to keep chemical cost down. This scenario suggests that it is possible for the optimal controller to sustain an increase in $\dot{V}_E$ without requiring any increase in $PaCO_2$ as long as the resulting gain in J_c outweighs the loss in J_m. Thus, in the absence of an exercise stimulus, the condition of isocapnic exercise hyperpnea is possible with the optimization model whereas this is not predicted by the chemoreflex model.

MODEL VALIDATION

The above qualitative arguments can be summarized by the following solution to the optimization equation. Substituting Eqs. (1)–(3) into Eq. (4) one obtains

$$J = \alpha^2 \left(P_I CO_2 + \frac{863\,\dot{V}CO_2}{\dot{V}_E(1 - V_D/V_T)} - \beta \right)^2 + 2\log \frac{\dot{V}_E}{(1 - \dot{V}_E/\dot{V}_{\max})} \tag{4a}$$

The minimum–cost solution is found by differentiating the above equation with respect to $\dot{V}_E$ and setting the derivative to zero. After some simplification the solution can be written as[5]:

$$\dot{V}_E = \frac{\dot{V}_{E_o}}{1 + \dot{V}_{E_o}/\dot{V}_{\max}} \tag{5}$$

where

$$\dot{V}_{E_o} = \alpha^2(P_a CO_2 - \beta)\frac{863\,\dot{V}CO_2}{(1 - V_D/V_T)} \tag{6}$$

Equations (5) and (6) define the optimal steady–state ventilatory response to CO_2 inhalation and exercise. The actual responses in $\dot{V}_E$ and $PaCO_2$ corresponding to any given P_ICO_2 and $\dot{V}CO_2$ are obtained by solving the simultaneous equations (1), (5) and (6). The variable $\dot{V}_{E_o}$ is the ideal ventilatory response in the absence of mechanical limitation (i.e., $\dot{V}_{\max} \longrightarrow \infty$). From Eq. (6) it can be seen that the ideal response is characterized by a linear $\dot{V}_E - PaCO_2$ relationship during CO_2 inhalation with constant $\dot{V}CO_2$ and a proportional $\dot{V}_E - \dot{V}CO_2$ relationship during exercise with resultant isocapnia. Furthermore, with combined CO_2 breathing and exercise the ideal ventilatory response exhibits a multiplicative interaction with respect to $PaCO_2$ and $\dot{V}CO_2$ (Fig. 3). This is in contrast to the reflex model which predicts an additive interaction (Fig. 3).

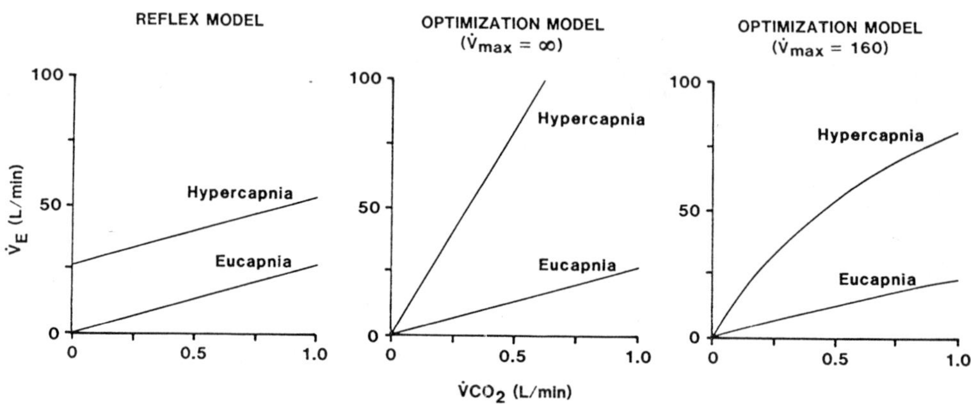

Fig. 3. Ventilatory responses to eucapnic and hypercapnic exercise (with constant $PaCO_2$) as predicted by the additive reflex model and the optimization model with and without mechanical limitation. (From Poon[5]).

The multiplicative effect predicted by the optimization model is in agreement with a similar synergistic effect of ventilatory CO_2–exercise interaction reported by some investigators but at variance with a purportedly additive interaction of these factors reported by others (see Ref. 7 for a thorough review). Thus, resolution of the actual form of ventilatory CO_2–exercise interaction is crucial for distinguishing the optimization and reflex models.

Such discrepancy in the reported effects of CO_2–exercise interaction may be largely explained by the variable influence of mechanical constraints on the ventilatory response. Because the system is expected to be mechanically limited at high $\dot{V}_E$, it is clear that any intrinsic interaction of stimuli would be significantly attenuated as ventilation builds up. It is therefore inappropriate to study ventilatory CO_2–exercise interaction without consideration of the operating levels of $\dot{V}_E$ and mechanical loads. If there is any multiplicative effect, it would most likely manifest itself at low rather than high $\dot{V}_E$. Furthermore, any such multiplicative effects would be greatly masked in the presence of significant ventilatory loads or in pulmonary disease.

The effect of mechanical limitation on the ventilatory response is modeled by the parameter $\dot{V}_{\max}$. From Fig. 4 which is a graphical illustration of Eq. (5), it can be seen that as the ideal ventilatory response approaches infinity the actual response is limited by a ceiling at $\dot{V}_{\max}$. Furthermore, the effect of mechanical limitation begins to exert its influence at relatively low ventilatory levels and assume an increasing predominance as ventilation rises. For example, at an ideal ventilatory response of $\dot{V}_{E_o} = \dot{V}_{\max}$ the actual response is predicted to reach only 50% of the ideal value.

The predicted effects of mechanical limitation on the exercise ventilatory response and CO_2–exercise interaction are illustrated in Fig. 3. In the control state where $\dot{V}_{\max} = 160 L/min$ is assumed, the model predicts an approximately linear $\dot{V}_E - \dot{V}CO_2$ relationship in eucapnia and a multiplicative CO_2–exercise interaction at low $\dot{V}_E$ levels in hypercapnic exercise similar to those in the unconstrained case. However, at higher $\dot{V}_E$ levels the hypercapnic exercise ventilatory response is seen to level off progressively reflecting the graded influence of mechanical limitation. Consequently, the initial multiplicative effect is gradually diminished while the additive effect becomes increasingly prominent. Extending this to even higher $\dot{V}_E$ levels (not shown) it is clear that as the ventilatory depression becomes more severe the multiplicative effects of CO_2 and exercise may even be reversed, resulting in a negative

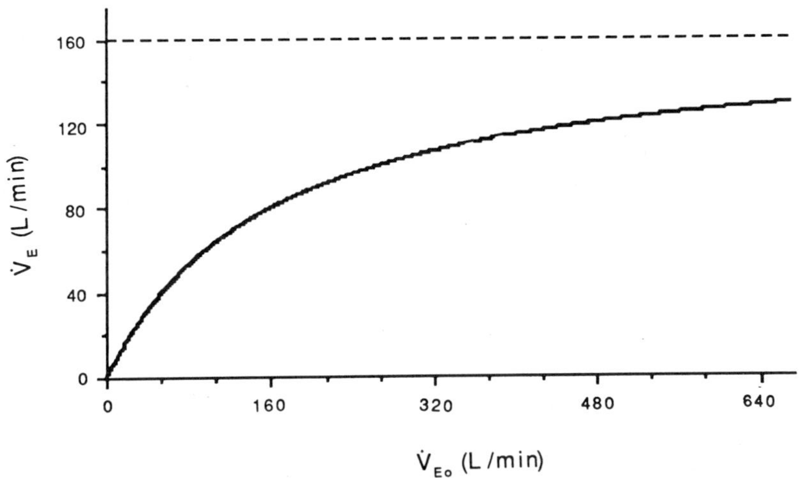

Fig. 4. Relationship between ideal and actual ventilatory response.

interaction. The variable nature of the predicted CO_2–exercise interaction at varying $\dot{V}_E$ levels may account for the rather disparate effects of exercise on the hypercapnic ventilatory response reported by different investigators.

The notion that mechanical factors may play a role in modulating the observed ventilatory CO_2–exercise interaction is further demonstrated by the predicted effects of mechanical loading on the exercise ventilatory response (Fig. 5). The presence of added mechanical loads is represented in the model by a lowering in $\dot{V}_{\max}$, which signifies the intensification of mechanical limitation. The magnitude of the simulated mechanical load is such that $\dot{V}_{\max}$ decreases by almost 40% from its control value. In the absence of any compensation for the load this will cause a similar reduction in $\dot{V}_E$. As shown in Fig. 5, during eucapnic exercise the ventilatory response under mechanical loading is almost completely restored to the unloaded values with only a slight depression in the slope of the $\dot{V}_E - \dot{V}CO_2$ relationship. In contrast, during hypercapnic exercise the ventilatory response is markedly depressed by mechanical loading, and CO_2–exercise interaction is considerably diminished.

To test the above theoretical predictions of the model we determined experimentally the effects of CO_2 inhalation, exercise and various mechanical loads on the ventilatory response in normal subjects. In order to enhance our ability to discern any multiplicative effects of CO_2 and exercise we confined the experimental range to mild or moderate exercise and hypercapnic levels. The experimental protocol of steady–state incremented exercise under a constant eucapnic or hypercapnic background (by end–tidal PCO_2 forcing) also allowed us to obtain more reliable CO_2–exercise interaction data in each subject than some conventional methods. As shown in Fig. 6 the observed effects of CO_2–exercise interaction resembled the theoretical predictions remarkably well during both unloaded breathing and inspiratory loading with an external resistance or elastance. The multiplicative effect of CO_2 and exercise was also verified in a separate study using a similar protocol but with arterial PCO_2 being directly measured from sampled blood[7].

The differing degrees of load compensation under various conditions may also be seen from the measured P_{100} response (Fig. 6). During eucapnic exercise, mechanical loading by IRL or IEL elicited a significant augmentation in P_{100} response in partial compensation for

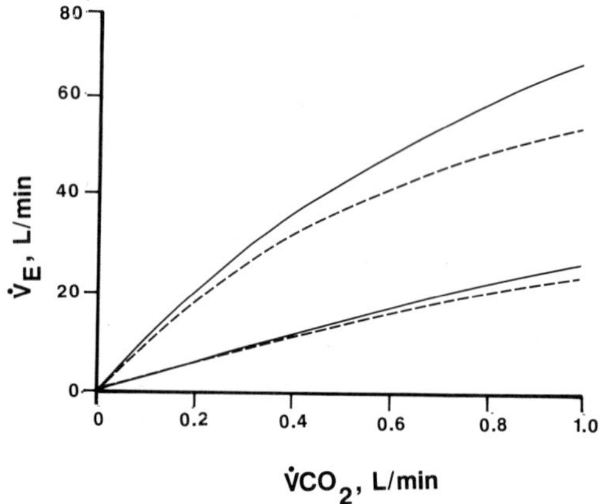

Fig. 5.   Predicted effect of mechanical loading (dashed lines) on ventilatory CO_2–exercise interaction. (From Poon[8])

the load. During hypercapnic exercise such augmentation in P_{100} was completely abolished under IEL, and although some compensatory response in P_{100} was seen under IRL, the percentage increase relative to the unloaded values was smaller than in the eucapnic exercise case. The decline in neurally mediated load–compensation response during hypercapnic exercise suggests that the decrease in ventilatory CO_2–exercise interaction under mechanical loading is not entirely a passive mechanical effect but may involve a control mechanism of central origin.

Not only was the neural compensatory response evoked during mechanical loading but a reverse pattern of load compensation during eucapnic exercise was also obtained when the intrinsic work of breathing was alleviated by means of mechanically assisted ventilation[11]. In the latter study, a reduction in the resistive work component was found to elicit a proportionate decrease in P_{100} response. Consequently, the initial hyperventilation caused by the ventilatory assist was largely corrected in the steady–state with nearly complete restoration of $\dot{V}_E$ and $P_{ET}CO_2$. Thus, the neural compensatory response was characteristic of eucapnic exercise whether the mechanical load was increased or decreased.

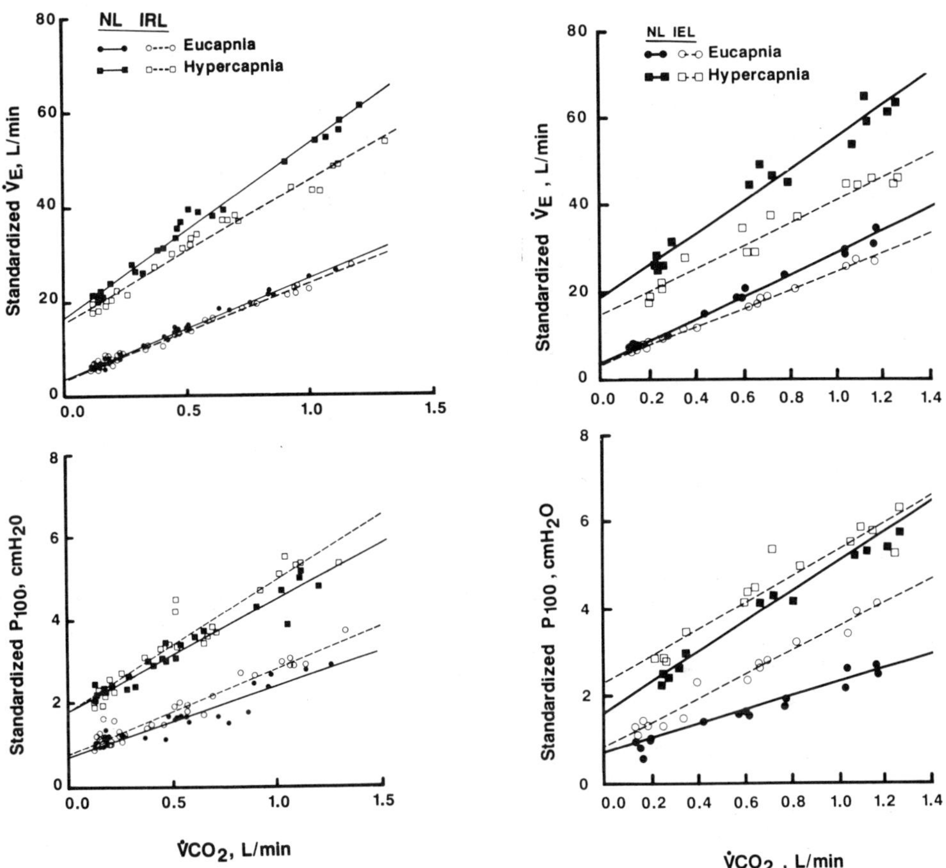

Fig. 6. Experimental data showing effects of various mechanical loads on ventilatory CO_2–exercise interaction. Data from different subjects were standardized using the method of Poon[10]. (From Poon[8,9])

There are two possible reasons for the relatively weaker load–compensation response observed during hypercapnic exercise. Firstly, with combined CO_2 and exercise inputs the operating level of $\dot{V}_E$ is already high and the effect of mechanical limitation is therefore likely to be more prominent than in eucapnic exercise. Secondly, it is possible that CO_2 breathing alone can impair the controller's ability to compensate for mechanical loads. To test the latter hypothesis, we compared in the same subjects the ventilatory exercise response and the CO_2 response at rest with and without added mechanical loads. As shown in Fig. 7 the hypercapnic ventilatory response was significantly lower in IRL or IEL than in the corresponding control states. The relative depression of the hypercapnic ventilatory response during mechanical loading was also significantly more severe than that found in eucapnic exercise over a similar range of $\dot{V}_E$ (cf. Fig. 6). The experimental results are in agreement with the previous finding that the ventilatory load–compensation response is generally more powerful in exercise than in CO_2 breathing[12].

The differential potency of the ventilatory load–compensation response under different experimental conditions may also be explained on the basis of the optimization model. During CO_2 breathing, there is little incentive for the controller to correct any hypoventilatory effect produced by mechanical loading because the inefficiency of pulmonary gas exchange makes it rather costly to do so. Such a degradation of pulmonary gas exchange function does not obtain during eucapnic exercise, however, and any substantial increase in chemical cost can be effectively avoided by increasing the mechanical output. Thus, the differential load-

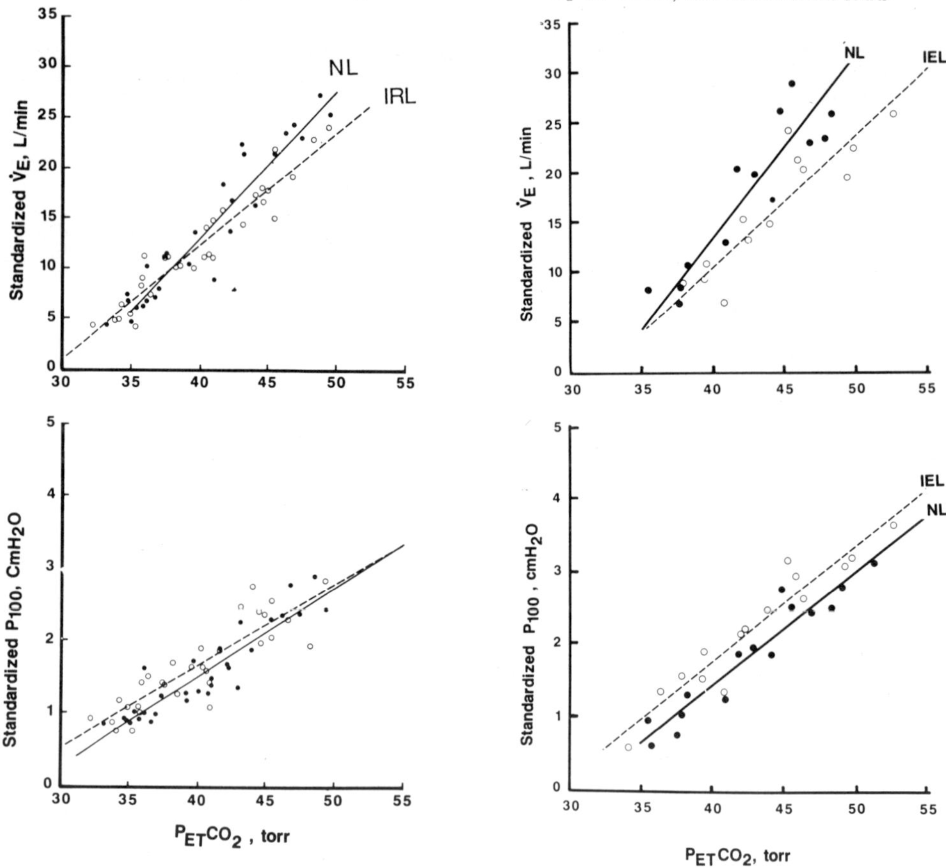

Fig. 7. Experimental data showing effects of various mechanical loads on hypercapnic ventilatory response.(From Poon[8,9])

compensation response during CO_2 breathing and exercise may stem from the same mechanism that differentiates the hypercapnic ventilatory response from the isocapnic exercise hyperpnea.

In summary, the optimization model successfully reproduces the following experimental findings: 1) a linear $\dot{V}_E - PaCO_2$ relationship during CO_2 inhalation; 2) an isocapnic $\dot{V}_E - \dot{V}CO_2$ relationship during exercise; 3) a multiplicative–type ventilatory CO_2–exercise interaction; 4) a depressant effect of mechanical limitation on CO_2–exercise interaction; 5) a neurally mediated compensatory response to ventilatory loading that is more pronounced during exercise than during CO_2 inhalation. Collectively, these experimental results are not mimicked by any other model.

MODELING PERSPECTIVES

What have we accomplished so far? Certainly the entire issue of respiratory control is still far from being settled; further research is needed to elucidate the precise control mechanisms underlying the various respiratory effects in health and in disease. Nevertheless, it is probably fair to say that indeed much has been gained from this modeling effort, and there are some emergent concepts that will prove useful in further understanding of respiratory control mechanisms.

Specifically, we have put in perspective the preeminent role traditionally accorded the simple reflex mechanism in describing respiratory control phenomena. While this remains a likely mechanism, we have demonstrated that it is by no means the only model that is capable of simulating ventilatory exercise and CO_2 response data. An optimization mechanism is equally – if not more – plausible. Indeed, we have shown that the latter has a number of interesting features that make it an attractive candidate for the control mechanism: it does not need to invoke any unknown "exercise stimulus", and it conforms to a wider range of experimental data including ventilatory CO_2–exercise interaction and load compensation. While these are not meant to constitute a definitive validation, they do offer a critical test for any alternative models that attempt to address this issue.

The most conclusive finding from this study is that the traditional model of a simple reflex controller with additive chemical and exercise inputs does not appear to be tenable under the above experimental test. Aside from the uncertainty of the postulated "exercise stimulus", this model fails to account for the characteristic interactions of CO_2, exercise, and mechanical loads in determining the ventilatory response. Although additional (yet unspecified) stimuli can always be invoked to explain such synergistic behavior, it is quite unlikely that the observed multiplicative effects of the various factors can be mimicked by a reflex controller in which all stimuli combine additively.

Given this background, it is tempting to assume that, for the basic reflex scheme to hold, a multiplicative process would have to exist within the controller to account for the CO_2–exercise interaction. This was, indeed, one of the options considered by Grodins in his classic paper 40 years ago[2]. Such a multiplicative reflex controller may even assume the same input–output characteristics (Eqs. (5),(6)) as the optimal controller and thus may exhibit the same stimulus interaction and load compensation properties. For example, it is conceivable that CO_2 sensitivity may be greatly enhanced during exercise through such a multiplicative controller and, consequently, the effect of mechanical loading may simply be compensated by a heightened chemoreflex mechanism, whereas such potentiation of the chemically mediated load–compensation mechanism is absent under resting conditions. The difference in the CO_2 sensitivity at rest and during exercise may, in turn, account for the differential potency of load compensation in CO_2 breathing and in exercise.

Although such an exclusively chemoreflex scheme may be theoretically plausible, it leaves open the important contribution of mechanosensory feedback which is widely recognized to play a significant role in mediating the load–compensation response[13]. Also pending clarification under this scheme is the physiological identity of not only the hypothetical exercise stimulus but, additionally, the complex multiplicative processing element within the controller.

Moreover, such a chemoreflex scheme offers no explanation for the seemingly optimal behavior in the control of breathing pattern. Thus, different control mechanisms must be invoked to describe the control of ventilation and breathing pattern. In contrast, we have shown that both these variables may theoretically be predicted simultaneously by use of a simple extension of the optimization model[14]. Such a generalization holds the possibility of a unified and coherent theoretical framework for describing both ventilatory and breathing pattern responses under a wide range of experimental conditions.

So, finally, here we are, starting from a relatively simple but daring hypothesis and then finding to our satisfaction that there might be some truth to it. There are, of course, many other questions that remain to be answered, and many other factors yet to be discovered. We have come a long way trying to discern the elements of chemoreflex, and it would probably take just at least as much to understand optimization. We have reached a point where we find ourselves at a historic crossroad that leads to new and exciting challenges and opportunities. The good news is that whichever course one chooses to take, one is bound to see new horizons given the benefit of the modeling perspective.

ACKNOWLEDGEMENT

This work was supported by NIH grants HL–30794 and HL–36239.

REFERENCES

1. C.S. Sherrington, "The Integrative Action of the Nervous System," Yale University Press, New Haven (1906).

2. F.S. Grodins, Analysis of factors concerned in regulation of breathing in exercise, Physiol. Rev. 30:220 (1950).

3. C.S. Poon, Optimal control of ventilation in hypoxia, hypercapnia and exercise, in: "Modelling and Control of Breathing," B.J. Whipp and D.M. Wiberg, eds., Elsevier, New York (1983).

4. F. Rohrer, Physiologie der Atembewegung, in: "Handbuch der normalen und path. Physiologie," A.T.M. Bethe et al., ed., Springer Verlag, Berlin (1925).

5. C.S. Poon, Ventilatory control in hypercapnia and exercise: optimization hypothesis, J. Appl. Physiol. 62:2447 (1987).

6. J.H. Milsum, "Biological Control Systems Analysis," McGraw–Hill, New York (1966).

7. C.S. Poon and J.G. Greene, Control of exercise hyperpnea during hypercapnia in humans, J. Appl. Physiol. 59:792 (1985).

8. C.S. Poon, Effects of inspiratory resistive load on respiratory control in hypercapnia and exercise, J. Appl. Physiol. 66:2391 (1989).

9. C.S. Poon, Effects of inspiratory elastic load on respiratory control in hypercapnia and exercise, J. Appl. Physiol. 66:2400 (1989).

10. C.S. Poon, Analysis of linear and mildly nonlinear relationships using pooled subject data, J. Appl. Physiol. 64:852 (1988).

11. C.S. Poon, S.A. Ward, and B.J. Whipp, Influence of inspiratory assistance on ventilatory control during moderate exercise, J. Appl. Physiol. 62:551 (1987).

12. N.R. Anthonisen, Some steady–state effects of respiratory loads, Chest 70, Suppl.: 168 (1976).

13. N.S. Cherniack and M.D. Altose, Respiratory responses to ventilatory loading, in: "Regulation of Breathing, Part II," T.F. Horbein, ed., Dekker, New York (1981).

14. C.–S. Poon, Optimality principle in respiratory control, Proc. Am. Control Conf. (1983).

CONSEQUENCES OF LUNG VOLUME OPTIMIZATION ON EXERCISE HYPERPNEA

Stanley M. Yamashiro and Fred S. Grodins

Biomedical Engineering Dept., University of Southern
California, Los Angeles, CA 90089-1451

INTRODUCTION

Previous work has shown that respiratory rate, airflow pattern shape, and the end-expiratory lung volume level can be predicted by an optimal control model based on minimum cycle work rate[1,2]. One prediction which can affect ventilatory control is a decrease in end-expiratory lung volume level which is graded according to the level of exercise. Such a decrease is consistently observed at the start of exercise and occurs in a feed-forward or predictive manner. One of the consequences of a decreased lung volume is lengthening of the diaphragm, which according to the length-tension characteristics of skeletal muscle increases active tension. Lowering of lung volume requires activation of expiratory muscles, and this work is stored as elastic energy and can be recovered during the following inspiration. This constitutes a potential additional ventilatory drive independent of chemical factors. Expiratory muscle work is stored as elastic work, and depending on whether activity is tonic or phasic will lead to different tidal volume increases. This paper makes a quantitative estimate of the magnitude of these effects on minute ventilation during exercise. Another consequence of lung volume optimization has to do with gas exchange. Lowering of lung volume could decrease dead space, which would increase alveolar ventilation for a given minute ventilation. However, a decrease in lung volume might also lead to collapse of airway units and increased heterogeniety of ventilation and perfusion which would impair gas exchange. Experiments dealing with this issue are also reviewed.

MATHEMATICAL MODEL

The respiratory muscle properties were described using the model of Younes and Riddle[3]. To this was added a non-linear relaxation pressure- volume curve that we previously used to explain lung volume shifts during exercise[2]. Frequency and end-expiratory lung volume were assumed to follow predictions of minimum cycle work rate[2]. To form a closed loop control model, a proportional controller considering CO_2 alone was assumed according to the formulation of Gray[4].

SIMULATION RESULTS

The effect of phasic expiratory muscle activation was simulated by applying 10 cm H_2O pressure during expiration only. Figure 1 shows the response in volume for this case.

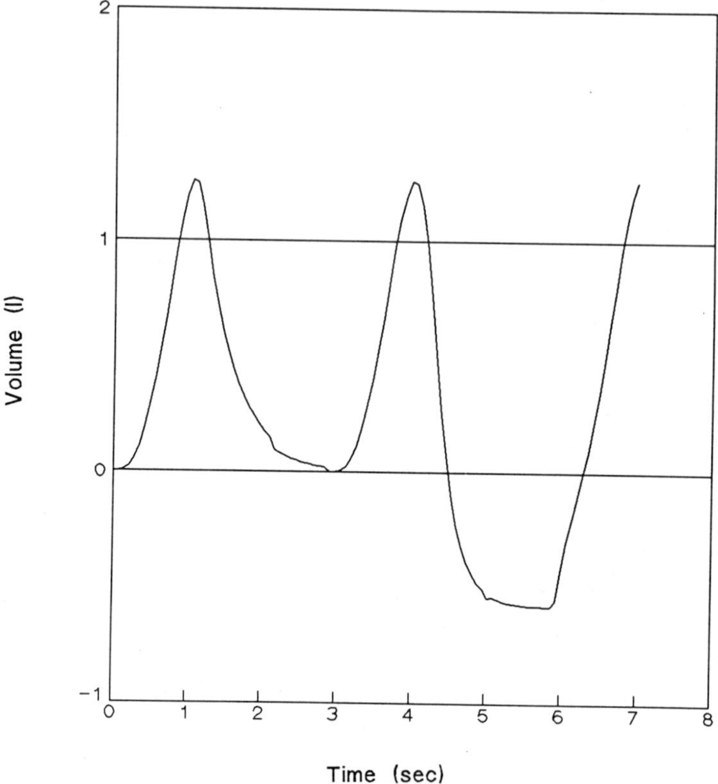

Fig. 1. Lung volume response to phasic expiratory muscle activation.

Lung volume is shown to shift below the baseline level by about 0.6 liter. The end-inspiratory lung volume level remains the same so tidal volume increases by the same magnitude as the reduction in end-expiratory lung volume,in this case by 0.6 liter. Figure 2 shows the tidal volume increase obtained by simulation for different levels of end-expiratory lung volume (or FRC-functional residual capacity) shifts for tonic (continuously applied pressure) and phasic expiratory muscle activation. Note that phasic activation results in a shift in end-expiratory lung volume equalling the tidal volume increase (100 % recovery) while tonic activation leads to only about a 20% recovery.

The consequence of phasic activation sufficient to optimize lung volume during exercise is shown in Figure 3. Shown in the figure are alveolar ventilation plotted versus metabolic rate for isocapnia (normal exercise response), proportional control (hypercapnia), and lung volume optimization. Proportional control leads to predictions consistently below the isocapnic level due to the expected rise in arterial CO_2. Lung volume optimization cannot be distinguished from isocapnia at low levels of metabolism (< 0.75 l/min), but eventually falls between predictions of isocapnia and proportional control. Even at fairly high levels of metabolism (1.25 = 5x resting), lung volume optimization explains half of the difference between isocapnia and proportional control.

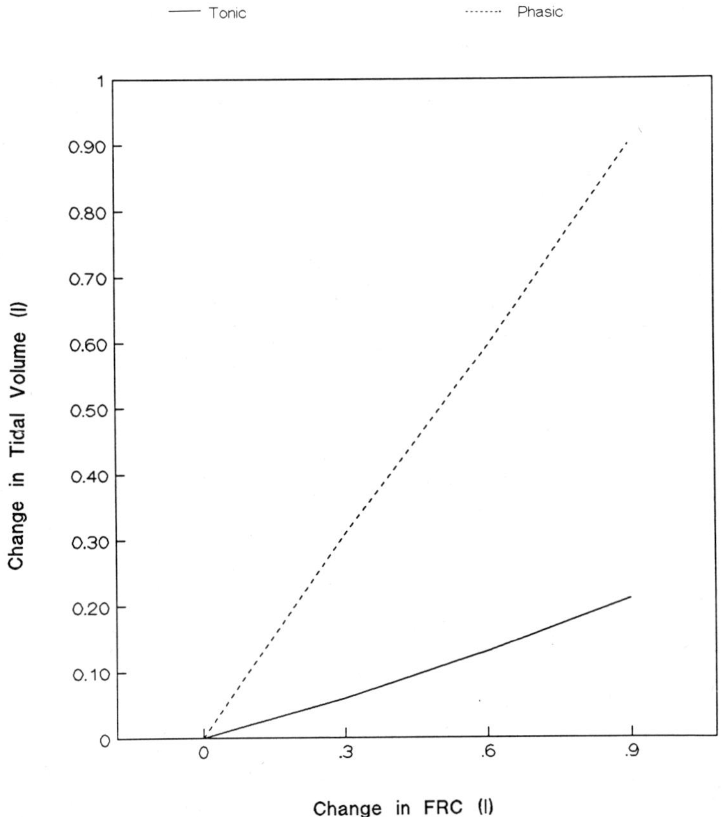

Fig. 2. Tidal volume increase for tonic and phasic muscle activation.

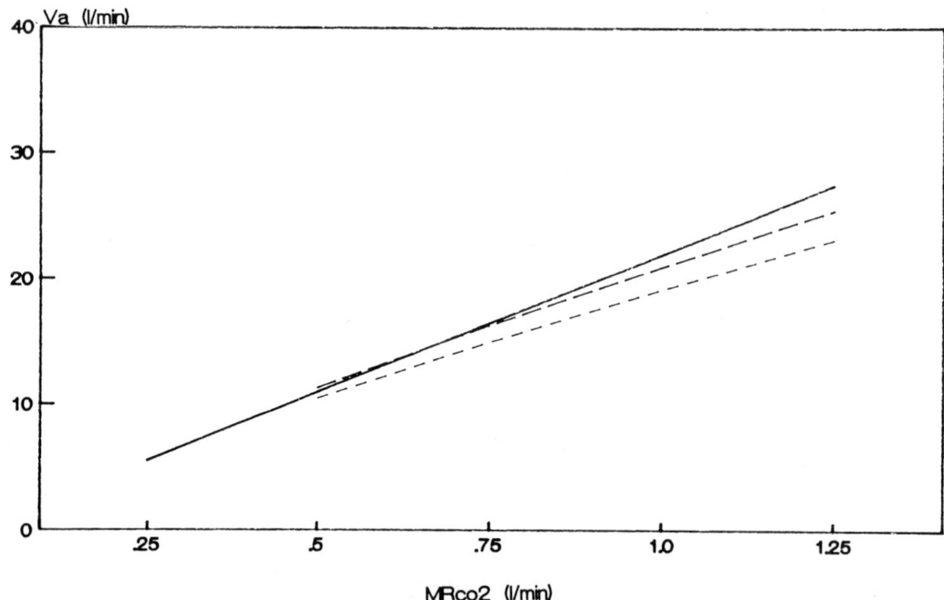

Fig. 3. Comparison of ventilatory responses for isocapnia, proportional control, and lung volume optimization.

ANIMAL EXPERIMENTS

The effects of applying a phasic change in lung volume was studied in six awake dogs by using negative end-expiratory pressure (NEEP). NEEP was applied at a level of -5 cm H_2O via a permanent tracheostomy during expiration only by using solenoid valves triggered by the airflow signal[5]. Figure 4 shows the mean ventilatory responses measured during NEEP application while breathing air or 4 % CO_2.

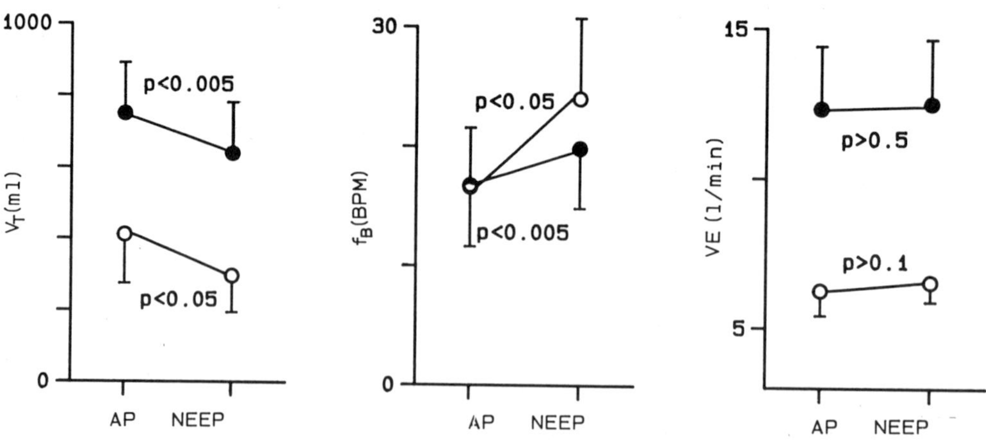

Fig. 4. Mean ventilatory responses to NEEP in 6 dogs.

42

Tidal volume fell and frequency rose during NEEP application while minute ventilation remained the same. Thus, the theoretical predictions of an increased tidal volume to phasically changed lung volume is not confirmed. Figure 5 shows the measured changes in arterial CO_2 tension and CO_2 response curve slope.

NEEP leads to a fall in $PaCO_2$ during air or CO_2 inhalation. There is also an increase in CO_2 response slope during NEEP. The change (NEEP-control) in arterial O_2 tension averaged 1.55 torr during air breathing (N=22, p < .05) and 1.24 torr during CO_2 inhalation (N=22, p < .05). The fall in $PaCO_2$ and rise in PaO_2 during NEEP with minute ventilation unchanged suggests an improvement in gas exchange, probably due to dead space reduction. Since PaO_2 improves, the expected deterioration of ventilation-perfusion heterogeniety due to a reduction in lung volume does not appear important.

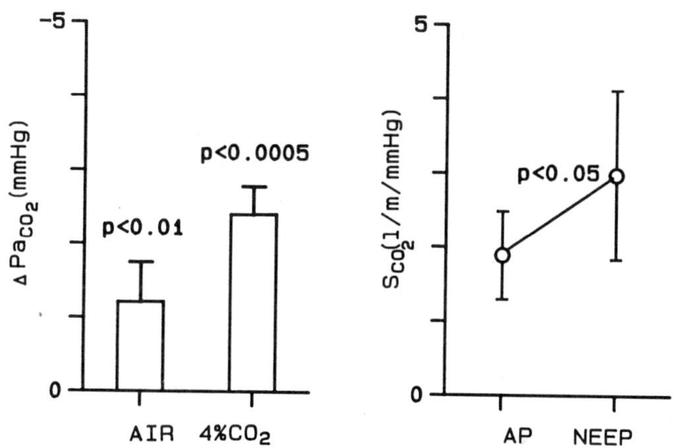

Fig. 5.   Arterial CO_2 and CO_2 response slope responses to NEEP. Changes in $PaCO_2$ produced by NEEP, breathing air or 4% CO_2. Changes in CO_2 respace slope produced by NEEP.

CONCLUSIONS

Theoretical predictions suggest that lung volume optimization mediated by phasic lung volume decreases during expiration can explain up to half the ventilatory response during moderate exercise. However, animal experiments where phasic lung volume changes are artificially imposed do not support this possibility. This negative result is probably due to the strong vagal inflation-deflation reflexes present in dogs which may not be a factor in man. Thus, the relevance of this mechanism to human responses remains unresolved. Animal experiments showed an unexpected improvement in gas exchange due to phasic reduction in lung volume. This may be an important factor involved in normal human exercise responses.

ACKNOWLEDGEMENTS

This study was supported by NIH Grants RR01861, HL16390, and HL33274.

REFERENCES

1. S.M. Yamashiro and F.S. Grodins, Optimal regulation of respiratory airflow, J. Appl. Physiol. 30:597 (1971).

2. S.M. Yamashiro and F.S. Grodins, Respiratory cycle optimization in exercise, J. Appl. Physiol. 35:522 (1973).

3. M. Younes and W.A. Riddle, Model for the relation between respiratory neural and mechanical outputs, I. Theory. 51:963 (1981).

4. J.S. Gray, The multiple factor theory of the control of respiratory ventilation, Science 102:739 (1946).

5. E.J. Cha and S.M. Yamashiro, Unpublished data.

EXPERIMENTAL DESIGN AND ANALYSIS FOR ASSESSING GAS EXCHANGE KINETICS DURING EXERCISE

George D. Swanson

Anesthesiology Dept., University of Colorado Medical School, Denver
and Physical Education Dept., California State University, Chico

INTRODUCTION

The study of pulmonary gas exchange kinetics during human exercise yields insight into control mechanisms involved in the process of delivery of oxygen from the environment to the exercising muscle and the consequent removal of carbon dioxide. Mathematical models play an important role in the design of these studies and for subsequent interpretation of gas exchange kinetic data. The modeling process provides for the interaction among experimental data, physiological interpretation and experimental design.

For our purposes here, it will be helpful to classify models with respect to intended purposes[1]. An empirical model summarizes observed behavior of a system with a model structure that may not necessarily be related to physiological structure. In contrast, a functional model attempts to relate physiological structure to observed behavior by summarizing the system behavior with a physiologically related structure. It differs from a descriptive type structural model (computer simulation) since it includes only structural aspects that are essential for describing the system response. The functional model is similar to the empirical model in that the model parameters are adjusted so that the model response mimics the system response.

In the following sections we shall illustrate these modeling concepts for assessing gas exchange kinetics during exercise.

ESTIMATING BREATH–BY–BREATH ALVEOLAR GAS EXCHANGE

Recent studies of dynamic–state pulmonary gas exchange utilize methods of estimating breath–by–breath alveolar gas exchange[2–5]. The standard algorithm utilizes the subject's resting FRC for the lung volume term[2]. However, a more descriptive algorithm for this purpose can be developed via the functional modeling approach. This algorithm yields estimates of breath–by–breath gas exchange and, as a by product of the functional model involved, yields estimates of effective lung volume and effective pulmonary blood flow[4,6].

A "first cut" functional model for the lung simply assumes a homogeneous gas mixing volume with a continuous pulmonary blood perfusion and a cyclic air perfusion as described by differential equations[7]. Analysis at the end of the expiratory time, yields a set of algebraic equations with the lung volume and pulmonary blood flow as unknown model parameters[6].

The parameter values for lung volume and pulmonary blood flow can be estimated so as to yield a minimal breath–to–breath variation in alveolar gas exchange[4,6,8–10]. These estimates are called effective lung volume (ELV) and effective pulmonary blood flow (EBF) because on a breath–to–breath basis the functional model represents the lung as if it were a homogeneous gas mixing chamber with a volume less than anatomical volume. The parameters represent that portion of the anatomical lung volume and blood flow that effectively participate in gas exchange on a breath–to–breath basis. This functional modeling approach yields an alveolar gas exchange extraction algorithm that can be implemented using matrix algebra techniques on personal computers.

SUBMAXIMAL PULMONARY GAS EXCHANGE KINETICS

Dynamic pulmonary gas exchange data reveal important aspects of oxygen transport and utilization for submaximal exercise. The O_2 consumption response data from a ten second pulse in work–rate exhibits two response modes[11]. The early mode reflects the rapid increase and decrease in cardiac output (cardiodynamic mode). The second mode reflects transport and utilization of O_2 at the muscular level.

The pulse work–rate input is useful because it mimics the ideal impulse used by systems engineers as an input for identifying physical systems[12]. The impulse has the property that it excites the system at all frequencies. These frequencies are of equal amplitude across the spectrum.

An alternative approach is to utilize sinusoidal inputs of equal amplitude at particular frequencies across the spectrum of interest. In this way, systems physiologists have utilized sinusoidal work–rate signals to asses the "quality" of the O_2 uptake system. The result is an amplitude and phase plot of the O_2 consumption response at given frequencies. For example, the average amplitude response (derived from PRBS data, see below) for six subjects in a study conducted by Stegemann et al. indicates a relative flat response in the frequency range of 1/450 to 1/75 cycles per second before bedrest[13]. In contrast, after seven days of bedrest, the "high" frequency response has been attenuated. This suggests that after bedrest, the O_2 transport/utilization system has been slowed and is unable to respond to "rapid" changes in work–rate.

Modeling of these data often utilizes an empirical model. For example, Bakker et al. have utilized this approach to characterize the O_2 consumption response to a sinusoidal work–rate input[14]. The frequency and phase response data were characterized by a four parameter empirical model including a series time delay parameter. In each of the four subjects, the model parameter values were estimated from the frequency response data. Although this model yields a good fit, for all four subjects, the time delay parameter value estimates were negative (–7.1, –8.4, –6.4, and –6.5 seconds respectively). A negative time delay can not be interpreted physiologically since it indicates that the output occurs before the input is applied! Thus, although this empirical model adequately summarizes the frequency response data, it can not be utilized for a physiological interpretation.

However, a more appropriate functional model can be deduced from the Bakker et al. data. This concept involves the relationship between amplitude and phase. That is, a minimum phase shift is required to achieve a given amplitude attenuation for physical systems characterized by a set of linear ordinary differential equations. This minimum phase shift can be determined by processing the amplitude response through a Hilbert transform[15–17]. If the measured phase shift is greater than the minimum phase, then a positive time delay in series with a dynamic model will characterize the data. However, with the Bakker et al. data, the measured phase shift is less than the minimum phase shift[15]. A series time delay estimate results in a negative number.

From these considerations, a parallel time delay two compartment model was chosen as a functional model candidate to characterize pulse data[11]. Six model parameters (gain, time delay and time constant parameter for each mode) can be estimated from the data.

The time delay parameters yield positive estimates. The two compartments are interpreted physiologically to correspond to the two response modes.

The cardiodynamic mode of the pulse response can be isolated by an appropriate input. Using a high frequency sinusoidal input, the slow mode response is attenuated leaving only the fast mode response.

The utility of this dynamic isolation technique is illustrated by a 30 second period sinusoidal work–rate response for both upright and supine exercise. The alveolar O_2 consumption peaks in phase with the sinusoidal work–rate for the upright position. In contrast, with the same subject pedaling the cycle ergometer from a supine position, the O_2 consumption response lags the sinusoidal work–rate[18]. In the upright position, action of the muscle pump yields rapid increases in mixed venous return and thus cardiac output. Therefore, an in phase O_2 uptake response is exhibited. In contrast, in the supine position, the blood volume is more evenly distributed throughout the body and thus, the muscle pump is not nearly as effective for increasing cardiac output temporally. The result is a lagging O_2 response.

DESIGN OF SUBMAXIMAL DYNAMIC WORK–RATE INPUTS

The two component response of the O_2 uptake system motivates specific design considerations for dynamic work–rate inputs. The first component is in response to an increase in cardiac output. This first early component is apparent with a pulse type input but is less apparent when a step change in work–rate is utilized as the forcing function, especially with work to work transitions[19-21]. Alternatively, a sinusoidal work–rate input function with a high frequency (30 sec period) can be utilized to isolate the cardiodynamic phase of the O_2 consumption response.

The second, slower phase of the response represents O_2 utilization and transport aspect of O_2 uptake. A variety of dynamic tests have been devised to yield estimates of this kinetic response. These dynamic tests offer an alternative to other non–invasive methods of assessing performance capacity, such as maximal oxygen uptake tests[13,22-26].

Most recently, investigators using this approach have utilized a pseudorandom binary sequence (PRBS) work–rate input for the study of O_2 consumption kinetics. In this type of exercise, the work–rate is switched in a random manner between two work–rate levels over a predefined period of time. For example, Stegemann and colleagues have examined oxygen consumption kinetics with a PRBS that had 15 segments, with 30 sec per segment for a total cycle period of 450 sec[13]. The work–rate was varied between two levels of power (20 watts and 80 watts), with a decision made by a computer algorithm every 30 sec whether to leave the work–rate at its existing level or to change to the alternative level.

The purpose of the Stegemann et al. study was to characterize O_2 consumption kinetics before and after seven days of bedrest. Using cross–correlation techniques and Fourier transform methods, a plot of frequency response amplitude and phase was used to assess the effect of the bedrest. The mean results for six subjects suggested that bedrest diminished the high frequency amplitude response. However, the data variability masked a precise statistical interpretation of the effect of seven day bedrest on a subject to subject basis.

Part of this variability is caused by the method the Stegemann group used to compute breath–by–breath O_2 consumption. The volume of O_2 inspired minus the volume of O_2 expired was used to compute VO_2 uptake at the mouth for a given breath (in–out algorithm). This method does not compensate for breath–to–breath changes in pulmonary stores and those changes add considerable noise to PRBS data. Under those conditions, it is difficult to assess the effects of interventions such as a short term bedrest because of the overriding presence of breath–to–breath noise.

We have recently introduced a direct approach to assessing the frequency response of a PRBS breath–by–breath O_2 consumption time series, which minimizes breath–to–breath noise[9,10]. Our approach combines the alveolar gas exchange algorithm with a Fourier series expansion over the PRBS period. This extraction algorithm approach yields estimates of the frequency response directly in terms of sinusoidal components and yields an estimate of the effective lung volume (ELV) for each subject. By design, our approach yields a minimal breath–to–breath variation due to fluctuations in pulmonary stores. For eight subjects, the average goodness of fit parameter (R^2) and the average standard error parameter (SE) on the Fourier coefficients were 0.37 and 28.0 ml/min respectively when the in–out algorithm was utilized. In contrast, using the extraction algorithm, $R^2 = 0.90$ and $SE = 6.1$. The advantage of this approach over the in–out algorithm approach is evident by considering the statistical interpretation of data from individual subjects[9].

The spectral content of the PRBS input can be designed to cover more of the "high" frequency region by reducing the minimum switch time from 30 sec. We have recently demonstrated the advantage of a 5 sec switch time PRBS over the 30 sec switch time PRBS in terms of high frequency signal content[27]. Alternatively, the optimal frequency content of an input work–rate signal can be designed specifically for parameter estimation and the frequency content can be translated to switching times of a binary (two state) input[28–30].

Thus, the design of the work–rate input can be optimized for parameter estimation given an appropriate functional model. Such functional models need further development and that development needs to be guided by available structural models[31]. However, the parallel time delay two compartment model has proven useful for a variety of studies where the work–rate input has sufficient dynamic pattern such as pulses, steps, PRBS and sine waves. Inputs of less dynamic nature, such as ramps, utilize a less descriptive model which often becomes empirical in structure. Consequently, ramp inputs yield poor estimates of physiological parameters related to the dynamic–state[32].

MAXIMAL WORK–RATE TESTS

Ramp tests to maximum oxygen consumption can be useful for detecting a "ventilatory threshold" when it exists or alternatively for characterizing curvilinear increases in carbon dioxide production relative to oxygen consumption. We have recently developed non–parametric models using smoothing splines for this purpose[33–35]. These models estimate the nominal time course of the time series data (univariate case) and its first and second derivative[34]. For characterizing breath–by–breath plots of CO_2 production versus O_2 consumption, a special case of the univariate algorithm can be utilized by rearranging the oxygen consumption data in an increasing order[35]. Alternatively, a bivariate algorithm can be used[33].

CONCLUSIONS

The experimental design and analysis for assessing gas exchange kinetics during exercise involves the use of models. In this paper, we have illustrated a variety of modeling concepts relevant to this task.

ACKNOWLEDGEMENTS

A similar paper was presented at the 36th annual meeting of the American College of Sports Medicine, Baltimore, 1989. The complete ACSM presentation is published elsewhere[36].

REFERENCES

1. G.D. Swanson, D.L. Sherrill and R.M. Engeman, Model utility in the study of cardio-respiratory control, <u>Ann. Biomed. Eng.</u> <u>11</u>:337 (1983).

2. W.L. Beaver, N. Lamarra, and K. Wasserman, Breath–by–breath measurement of true alveolar gas exchange, J. Appl. Physiol. 51:1662 (1981).

3. N. Lamarra, B.J. Whipp, S.A. Ward, D.J. Huntsman, and K. Wasserman, Effect of inter–breath fluctuations on characterizing exercise gas–exchange kinetics, J. Appl. Physiol. 62:2003 (1987).

4. G.D. Swanson, Breath–to–breath considerations for gas exchange kinetics, in: "Exercise, Bioenergetics and Gas Exchange," edited by P. Cerretelli and B.J. Whipp. Amsterdam: Elsevier/North Holland, p. 211 (1980).

5. G.D. Swanson and D.L. Sherrill, A model evaluation of estimates of breath–by–breath alveolar gas exchange, J. Appl. Physiol. 55:1936 (1983).

6. D.L. Sherrill, B.H. Dietrich and G.D. Swanson, On the estimation of pulmonary blood flow from CO_2 production time series, Comput. Biomed. Res. 21:503 (1988).

7. K.H. Weisiger and G.D. Swanson, Importance of oscillations in alveolar gas concentrations in the analysis of rebreathing data, J. Appl. Physiol. 61:1104 (1986).

8. D.L. Sherrill and G.D. Swanson, Application of the general linear model for smoothing gas exchange data, Comput. Biomed. Res. 22:270 (1989).

9. G.D. Swanson, R.G. May, B.H. Dietrich and R.L. Hughson, Assessing breath–by–breath pseudorandom binary sequence O_2 consumption time series, J. Appl. Physiol. submitted (1989).

10. G.D. Swanson, R.G. May, B.H. Dietrich and R.L. Hughson, Characterizing breath–by–breath pseudorandom binary sequence gas exchange time series, Comput. Biomed. Res. submitted (1989).

11. R.L. Hughson, D.L. Sherrill and G.D. Swanson, Kinetics of VO_2 with impulse and step exercise in man, J. Appl. Physiol. 64:451 (1988).

12. D.L. Sherrill, N.B. Kindig, G.D. Swanson and R.W. Virtue, Analysis of dynamic cardio-respiratory response to pulse work load, in: "Modelling and Control of Breathing," edited by B.J. Whipp and D.M. Wiberg. New York: Elsevier Biomedical, p. 237 (1983).

13. J. Stegemann, D. Essfeld and U. Hoffmann, Effects of a 7–day head–down tile (−6°) on the dynamics of oxygen uptake and heart rate adjustment in upright exercise, Aviat. Space Environ. Med. 56:410 (1985).

14. H.K. Bakker, R.S. Struikenkamp and G.A. Devries, Dynamics of ventilation, heart rate, and gas exchange: sinusoidal and impulse work loads in man, J. Appl. Physiol. 48:289 (1980).

15. D.L. Sherrill, D.L. and G.D. Swanson, Minimum phase considerations in the analysis of sinusoidal work, IEEE Trans. Biomed. Eng. 28:832 (1981).

16. G.D. Swanson and D.L. Sherrill, Application of a Hilbert transform for system identifi–cation in respiratory control studies, IEEE Trans. Syst. Man Cybern. 12:582 (1982).

17. D.F. Tuttle, "Electric Networks," New York: McGraw–Hill, (1965).

18. G.D. Swanson, Exercise strategies and assessment of cardiorespiratory fitness in space. in: "The Case for Mars III," San Diego: Univelt Press, in press (1989).

19. D. Linnarsson, Dynamics of pulmonary gas exchange and heart rate changes at start and end of exercise, Acta Physiol. Scand. Suppl. 415:1 (1974).

20. O. Wigertz, Dynamics of respiratory and circulatory adaptation to muscular exercise in man. A systems analysis approach, Acta Physiol. Scand. Suppl. 363:1 (1971).

21. B.J. Whipp, S.A. Ward, N. Lamarra, J.A. Davis and K. Wasserman, Parameters of ventilatory and gas exchange dynamics during exercise, J. Appl. Physiol. 52:1506 (1982).

22. P. Cerretelli, D. Pendergast, W.C. Paganelli and D.W. Rennie, Effects of specific muscle training on $\dot{V}O_2$ on–response and early blood lactate, J. Appl. Physiol. 47:761 (1979).

23. D. Essfeld, U. Hoffman and J. Stegemann, VO_2 kinetics in subjects differing in aerobic capacity: investigation by spectral analysis, Eur. J. Appl. Physiol. 56: 508 (1987).

24. R.C. Hickson, H.A. Bomze and J.O. Holloszy, Faster adjustment of O_2 uptake to the energy requirement of exercise in the trained state, J. Appl. Physiol. 44:877 (1978).

25. R.L. Hughson, Alterations in the oxygen deficit–oxygen debt relationships with beta-adrenergic receptor blockade in man, J. Physiol. (London) 349:375 (1984).

26. P.C. Murphy, L.A. Cuervo and R.L. Hughson, Comparison of ramp and step exercise protocols during hypoxic exercise in man, Cardiovas. Res. in press (1989).

27. R.L. Hughson, D.A. Winter, A.E. Patla, G.D. Swanson and L. A. Cuervo, Investigation of VO_2 kinetics in man with pseudorandom binary sequence work rate change, J. Appl. Physiol. submitted (1989).

28. R.M. Engeman, G.D. Swanson and R.H. Jones, Sinusoidal frequency allocation for estimating model parameters, IEEE Trans. Syst. Man Cybern. 11:243 (1981).

29. R.M. Engeman, G.D. Swanson and R.H. Jones, Optimal frequency locations for estimating model parameters in studies on respiratory control, Comput. Biomed. Res. 16:531 (1983).

30. R.M. Engeman, G.D. Swanson and R.H. Jones, Input design for model discrimination with application to respiratory control during exercise, IEEE Trans. Biomed. Eng. 26:579 (1979).

31. T.J. Barstow and P.A. Mole, Simulation of pulmonary O_2 uptake during exercise transients, J. Appl. Physiol. 63:2253 (1987).

32. G.D. Swanson and R.L. Hughson, On the modeling and interpretation of oxygen uptake kinetics from ramp work rate tests, J. Appl. Physiol. 65:2453 (1988).

33. S.J. Anderson, R.H. Jones and G.D. Swanson, Smoothing polynomial splines for bivariate data, SIAM J. Sci. Stat. Comput. in press (1989).

34. T.D. Wade, S.J. Anderson, J. Bundy, V.A. Ramadevi, R.H. Jones and G.D. Swanson, Using smoothing splines to make inferences about the shape of gas–exchange curves, Comput. Biomed. Res. 21:16 (1988).

35. D.L. Sherrill, S.J. Anderson and G.D. Swanson, Using smoothing splines for detecting ventilatory thresholds, Med. Sci. Sports Exercise in press (1989).

36. G.D. Swanson, Assembling control models from pulmonary gas exchange dynamics, Med. Sci. Sports Exercise in press (1989).

Part II:

CHEMICAL CONTROL OF VENTILATION – EMPIRICAL MODELS AND PARAMETER ESTIMATION

PARAMETER ESTIMATION THEORY FOR RESPIRATORY PHYSIOLOGY

Donald M. Wiberg

Department of Electrical Engineering
Department of Anesthesiology
University of California, Los Angeles, CA   90024-1594

The Belgian painter, Rene´ Magritte, painted "The Betrayal of Images " in 1929, in which a realistic painting of a pipe is over the statement "This is not a pipe". His very graphic point is that the image of a pipe is not reality. In direct analogy, it must be remembered that the mathematical equations describing a physiological process are not reality, either. An equation such as $PV = RT$ is only an approximation to nature, and does not hold for shock waves, very high pressures, very low pressures, etc. If we were to measure the pressure-volume relationship of a gas at a constant temperature, the measurements could be made only to a certain accuracy. The point is that $PV = RT$ is a useful approximation for many purposes for adequately describing the behavior of a gas, and as such is *robust*. However, choice of this model $PV = RT$ depends on the purpose for which the model is to be used.

Models are either dynamic or static, depending upon whether the behavior with time is considered or not. Linear static models are those such as

$$\dot{V} = m\, P_{ETCO_2} + b \tag{1}$$

and nonlinear static models are those such as the inclusion of the "dog's leg" shape of the above linear model. Dynamic models of linear systems can be described by transfer functions, such as that between changes in ventilation and changes in end tidal CO_2 partial pressure:

$$\dot{V}(s) = \left[\frac{g_1}{\tau_1 s + 1} e^{-sT_1} + \frac{g_2}{\tau_2 s + 1} e^{-sT_2} \right] P_{ETCO_2}(s) \tag{2}$$

whereas nonlinear dynamic models cannot be expressed in such form, even when the nonlinearities are slight such as τ_2 depending upon P .

In summary, regarding models, the purpose of the model is primary in system identification. Model structure must be rich enough to describe expected behavior. However, the model structure must be simple enough to give confidence in individual parameter values. If too many parameters are in the model, the data may not give enough confidence in some of the parameter values.

Models can be classified as to static or dynamic. Here only dynamic models will be considered. Linear, time-invariant models are usually assumed. However, these assumptions should be tested before proceeding with parameter estimation. For example, stimuli (inputs) may be administered in a sinusoidal manner, until the responses (outputs) assume a sinusoidal shape (i.e. until transients have died away). Then the experiment can be repeated with the input at two and/or three times the amplitude used in the first test. If the outputs are not correspondingly two and/or three times their original amplitude, the system is not linear. This should be done over the whole range of frequencies of interest, as determined by the purpose of the model. Furthermore, some of the experiments should be repeated on different days to determine the time-variability of the model. Finally, time delays of the model should be ascertained by inputs that are pulses or steps. The time from the administration of the input until the first reaction by the output is the time delay.

Many models can be based upon physiological principles. Especially useful is the retention of physiologically meaningful parameters to be estimated. It is only mathematically convenient to have a model that is linear in the parameters, such as a and b in

$$\frac{1}{s^2 + a\,s + b} \tag{2}$$

but it may often be more useful and suitable to have a model that is nonlinear in the parameters, such as the poles p_1 and p_2 in

$$\frac{1}{(s + p_1)(s + p_2)} \tag{3}$$

If the model is time-invariant, linear, and linear in the parameters, it can be put into one of several standard types of model. Often systems are assumed to be of such a standard type if there is no other theory to support choice of another model. This technique is called "black-box" modeling. It can be very instructive to compare the results of black box modeling with the results of physiological modeling, because errors and unwarranted assumptions can be uncovered in the physiological model. Furthermore, system identification packages for the computer are available to quickly perform black box modeling, so the cost in time and effort is minimal. Of course, the physiological model can inherently build in fewer and more meaningful parameters than the black box model.

Standard black box models involve the stimulus (input) u, the response (output) y, the white noise w, and s the Laplace transform variable. If $A(s)$, $B(s)$, and $C(s)$ are polynomials in s (polynomial valued matrices in the multiple-input, multiple-output case), then the autoregressive (AR) model is defined by

$$A(s)\,y = w \tag{4}$$

The moving average (MA) model is defined by

$$y = C(s)\,w \tag{5}$$

The autoregressive model with exogenous input (ARX) is defined by

$$A(s)\,y = B(s)\,u + w \tag{6}$$

The autoregressive, moving average model (ARMA) is defined by

$$A(s)\,y = C(s)\,w \tag{7}$$

Finally, the autoregressive, moving average model with exogenous input (ARMAX) is defined by

$$A(s)\,y = B(s)\,u + C(s)\,w \tag{8}$$

For discrete time systems, the s variable is replaced by q^{-1}, the unit delay operator.

All the above models can be put in the linear state space form by appropriately defining a state vector $\vec{x}(t)$. Then the linear state space model is

$$d\vec{x}/dt = A\vec{x} + B\vec{u} + \vec{w}_1 \tag{9}$$

$$\vec{y} = C\vec{x} + \vec{w}_2 \tag{10}$$

where A, B, and C are matrices of appropriate dimension and $\vec{w}_1$ and $\vec{w}_2$ are vector valued white noise processes.

There are no corresponding forms for nonlinear systems. A general nonlinear function $\vec{f}$ must be found by theory or experiment. If the nonlinear model is independent of input and time, it is autonomous:

$$d\vec{x}/dt = \vec{f}(\vec{x}) \tag{11}$$

If the model can be influenced by the input, then it is time-invariant:

$$d\vec{x}/dt = \vec{f}(\vec{x},\vec{u}) \tag{12}$$

If the system also depends on the time at which the experiment is performed, then the model is non-linear and time-varying:

$$d\vec{x}/dt = \vec{f}(t,\vec{x},\vec{u}) \tag{13}$$

There are many other mathematical expressions for models: partial differential, stochastic, etc., but the above are the most commonly used. These models are generally parameterized by an unknown parameter vector θ, which is to be estimated by fitting to measured input-output pairs. Before considering the parameter estimation problem, the identifiability and sensitivity of the model should be investigated.

For the static model

$$y = f(u;\theta) \tag{14}$$

the sensitivity $\partial y/\partial\theta$ is defined as

$$\frac{\partial y}{\partial\theta} = \frac{\partial f}{\partial u}(u;\theta) \tag{15}$$

which depends on the input u in addition to θ. Since θ is unknown, the sensitivity must be evaluated for a range of possible parameter values.

For the dynamic model with no noise,

$$dy/dt = f(t,y,u;\theta) \qquad y(0) = y_o \tag{16}$$

the output y is the mathematical expression $\psi(t;y_o,u_{(0,t)},\theta)$ that solves the above expression (16), where ψ is called the trajectory and depends not only on t but the initial conditions, the input values over the time interval from 0 to t, and the parameter θ. Analogous to the static model case, the sensitivity is defined as

$$\frac{\partial y}{\partial\theta} = \frac{\partial\psi}{\partial\theta}(t;y_o,u_{(0,t)},\theta) \ . \tag{17}$$

The no noise dynamic model is not identifiable when there exists some non-zero constant vector α such that

$$\alpha^T\frac{\partial\psi}{\partial\theta}(t;y_0,u_{(0,t)},\theta) = 0 \tag{18}$$

for all t. Specifically, if there is only one parameter so that θ is one dimensional, the above equation reduces to zero sensitivity. For zero sensitivity, any change in θ can not be seen at the output y, hence the relation between sensitivity and identifiability.

When the model also models noise, the system is not identifiable when the left side of (18) above is indistinguishable from the noise. Furthermore, for a given amount of noise, the larger the sensitivity $\partial\psi/\partial\theta$, the more accurately θ can be determined from measurements of the output y. Conversely, the larger the sensitivity $\partial\psi/\partial\theta$, the same error in θ will give larger errors in predicting the output y.

Computation of the sensitivity will give a rough estimate of the parameter variance before doing any experiments. The rough estimate is

$$\mathrm{var}(\theta) \approx \frac{\mathrm{var}(w)}{N} \left[\frac{1}{t_f} \sum_{t=1}^{t_f} \frac{\partial \psi}{\partial \theta}(t) \frac{\partial \psi^T}{\partial \theta}(t) \right]^{-1} \qquad (19)$$

The number N is the number of data points, where $N = t_f$ if the dimension of θ is one, and the quantity in brackets is the time average of the square of the sensitivity. Also, here $\mathrm{var}(w)$ should be taken to be an estimate of the variance of all the uncertainty in the system, not only due to measurements but also mismodelling, etc.

For example, for known α, the system

$$y(t) = \alpha \theta + w(t) \qquad (20)$$

has sensitivity $\partial \psi / \partial \theta = \alpha$ so that the rough estimate (19) is

$$\mathrm{var}(\theta) = t_f^{-1} \alpha^{-2} \mathrm{var}(w) \qquad (21)$$

which is exact in this case if w is white and the model (20) is exact.

The identification procedure can now be summarized as follows. First, decide on the model, on the basis of purpose and robustness. Second, parameterize the model. Third, find the sensitivity. Fourth, roughly estimate the parameter variance. Fifth, go back to one if the parameter variance is unacceptable. Otherwise, proceed. Sixth, do simple experiments to see if the system is of the correct model form, such as linear. Seventh, only if the preceding is satisfactory, then parameter estimation should be performed from detailed experiments.

Only at this point is it proper to consider the method of estimation of parameters. If the model is nonlinear, the only alternatives are least squares or forms of series solution such as Wiener kernels. Therefore only choices of parameter estimation techniques will be discussed for linear systems.

Here only off-line estimation will be discussed, i.e. all the data is collected before parameter estimation begins. Otherwise see reference 3.

Experimental time length must be greater than ten data points per parameter, otherwise the asymptotic rules assumed here do not apply.

The data sampling interval should be about ten times the anticipated system bandwidth. If the sampling is too fast, round off errors accumulate and bad fits at high frequency result. If the sampling is too slow, aliasing results.

The input should be persistently exciting, so that longer data records will improve the estimation accuracy. Persistently exciting means the input has enough amplitude in the frequency range of interest to determine the parameter values. For example, white noise through any system is persistently exciting, but a constant input cannot determine more than one parameter value. Constant P_{ETCO_2} has been shown ineffective at determining both central and peripheral chemoreceptor gains. Our laboratory has shown that one up and one down step in P_{ETCO_2} will adequately estimate both gains, which is enough excitation for this case. However, the amount of persistency of excitation needed is determined by experimental conditions.

The best parameter estimation technique for time-invariant linear systems is to use a number of sinusoidal inputs over the frequency domain of interest. Then the frequency response of the transfer function can be estimated directly. Unfortunately this technique often cannot be used, either because it is impractical to generate sinusoids in the input or because the experimental time is not long enough.

If frequency domain methods cannot be used, then either prediction error techniques, such as least squares and maximum likelihood, or correlation of prediction error and past residuals, such as instrumental variables, should be used. A discussion of these techniques is given in references 1, 2, and 4. This is a complicated subject, and there is not adequate space available here for a discussion other than it is probably best to begin with least squares.

Before processing the data through any parameter estimation algorithm, both input and output should be pretreated in two ways. First, the data should be visually scanned or run through an outlier detection algorithm to eliminate false data, data gaps, and inappropriate or impossible data. Second, both input and output should be run through the same prefilter. This prefilter can be designed using the techniques found in chapter 8, section 5 of reference 1. In particular, it is shown that ARX models weight the high frequency much too much. Therefore the prefilter must be low pass for an ARX model.

Once the parameters are estimated, the job is NOT finished! The model with the estimated parameters must be tested against reality. First, estimates of the parameter variance must be computed, usually from the Cramer-Rao lower bound. Second, compute the residual $\varepsilon(t) = y(t) - \hat{y}(t; \theta^*)$, i.e. system output minus output from the model with the estimated parameters. The autocorrelation and spectrum of ε should be plotted. The plots must show ε *almost* white in the frequency range of interest. (No real system gives truly white ε.) Third, plots of the cross correlation and spectrum of ε and u must also be almost white. Fourth, the probability distribution of ε should be plotted from a histogram if ε is almost white. Any bimodality or other distribution far from Gaussian should be investigated. Fifth, the no noise model with the input u and parameter value θ^* must be compared with the true output $y(t)$. Sometimes the estimator can predict the noise rather than estimate the parameters, or local minima in the cost function are attained, rather than the global minimum. Our laboratory has examples where all other tests were passed, but not this fifth one.

Finally, all the tests and all the model validation should be done for models of different order. Appropriate model orders can be estimated by the Akaike information criterion (AIC), ref. 5. The AIC has been shown asymptotically equivalent to the likelihood ratio test, pole-zero cancellation, and the F-test. However, the AIC test is NOT consistent, and tends to overestimate the system order. The minimum description length principle of Rissanen is consistent and gives an unbiased measure, of which there are many. However, this principle gives the accumulated prediction error criterion for ARX systems, ref. 6. The accumulated prediction error criterion was adapted to on-line use by Valcke in his UCLA Ph.D. thesis. He used this to detect the gradual appearance of another compartment, such as an edema in the lung.

In conclusion, the modeling, identification, and parameter estimation of any physical or physiological process is an iterative and interactive procedure. Simple models based on physical or physiological reasoning must be compared with experimental data. Often this comparison will lead to revision of the model and the redesign of the experiment. Throughout this iterative procedure the purpose of the model must be kept in mind. The theory of system identification has progressed rapidly in the past ten years, and much of this progress was caused by the inability of the previous theory to explain experimental data.

References

1. Lennart Ljung, *System Identification Theory for the User*, Prentice Hall, New Jersey, 1987.

2. Torsten Söderström and Petre Stoica, *System Identification*, Prentice Hall International, Hemel Hempstead, England, 1988.

3. Lennart Ljung and Torsten Söderström, *Theory and Practice of Recursive Identification*, MIT Press, Cambridge, Mass., 1983 (paperback 1987).

4. Graham C. Goodwin and Robert L. Payne, *Dynamic System Identification: Experiment Design and Data Analysis*, Academic Press, New York, 1977.

5. Akaike, H. "A new look at statistical model identification", *IEEE Trans. Auto. Control*, AC-19, 6, pp. 716-723, Dec. 1974.

6. Rissanen, J., "Predictive least squares principle", Research report p.1-21, IBM Research, San Jose, CA, 1984.

DYNAMIC END-TIDAL FORCING TECHNIQUE: MODELLING THE
VENTILATORY RESPONSE TO CARBON DIOXIDE

Jacob DeGoede and Adriaan Berkenbosch

Department of Physiology, University of Leiden
P.O.Box 9604, 2300 RC Leiden, The Netherlands

INTRODUCTION

The chemical regulation of breathing, the primary purpose of which is to ensure CO_2, O_2 and H^+ homeostatis in arterial blood and brain tissue, involves two major chemosensitive structures, the central chemoreceptors and the peripheral chemoreceptors. In spite of much experimental work a definite answer to the question of the relative importance of the peripheral and the central chemoreceptors in the control of breathing is still lacking. Different classes of techniques have been used to separate the contributions of these receptors to the ventilatory responses following changes in blood gas tensions and acid-base disturbances (see ref. 1 and references cited therein). Among the different classes of techniques dynamic analysis methods are particularly attractive because they can be applied non invasively. The dynamic end-tidal forcing (DEF) technique, developed by Swanson and Bellville in 1970's, is perhaps the most promising. This technique consists of three elements: 1) accurate breath-to-breath control of the end-tidal CO_2 and O_2, so that they can be forced to follow a prescribed pattern in time together with the measurement of the breath-to-breath ventilation; 2) mathematical modelling of the input-output relationship of the respiratory controller; 3) techniques to estimate the model parameters from the noisy input-output data.

In this article we will not discuss the theoretical and practical aspects of the control of the end-tidal gas concentrations. Furthermore, we restrict ourselves to the ventilatory response following isoxic changes in P_{ET,CO_2}. For a general review of models in the control of breathing we refer to Khoo and Yamashiro[2]. A summary of the work on the DEF technique by Bellville and co-workers can be found in ref. 3.

MATHEMATICAL MODELS

In modelling a system there is always a trade-off between the complexity of the model and the number of model parameters which can be meaningfully estimated from the data obtained in an experiment. Since there are two main groups of chemoreceptors it is to be expected that the dynamic description of the effects of isoxic changes in P_{ET,CO_2} has to contain at least two CO_2 related ventilatory components. The additive two-compartment model, introduced by Swanson and Bellville[4] and subsequently modified by Wiberg et al.[5] seems to describe the experimental data adequately in a variety of experimental situations and the number of parameters is small enough for parameter estimation applications. In this section we will describe this model in detail.

Since the breath-to-breath noise in the experimental data is one of the major obstacles in the extraction of physiological parameter values we will pay special attention to the modelling of the noise. Since the measurements are made on a breath-to-breath basis the model equations are written in discrete time, although a continuous time representation is usually presented for the sake of simplicity (e.g. ref. 6). We will use a state space representation and write the model equation in matrix notation. From now on we restrict ourselves to square wave input signals, unless otherwise stated. We introduce the state vector

$$X(n) = \{X_1, X_2, X_3\}^\top, \tag{1}$$

in which X_1 is the central ventilatory component, X_2 the peripheral ventilatory component and X_3 an external noise component. The symbol $\top$ denote "transpose". The additive two-compartment model in state space form is

$$X(n+1) = A(n+1)X(n) + B(n+1)U(n+1) + DV(n), \tag{2}$$

with output function

$$Y(n) = CX(n) + ct(n) + W(n). \tag{3}$$

The output $Y(n)$ is the ventilation of the nth breath, which occurs at the time $t(n)$. In eqn (3) $W(n)$ denotes measurement noise and c is a drift parameter. The system matrix $A(n)$ is a diagonal matrix with

$$\alpha_{ii}(n) = \exp\{-\alpha_i(n)\Delta t(n)\}(i = 1,\ 2,\ 3), \tag{4}$$

where

$$\alpha_1(n) = mP_{ET,CO_2}\,(t(n) - T_1) + b \tag{5}$$

and α_2 and α_3 constants. The length of the nth breath is $\Delta t(n)$ and T_1 the central time delay i.e. the time needed to carry a P_{ET,CO_2} disturbance from the lungs to the central chemoreceptors. In eqn (5) the constants m and b allow for the possibilty that the time constant of the central ventilatory on-transient (τ_{on}) is different from that of the off-transient (τ_{off}). The distribution matrix $B(n)$ is a diagonal matrix with elements

$$\beta_{11} = g_1[1 - \alpha_{11}], \tag{6}$$

$$\beta_{22} = g_2[1 - \alpha_{22}] \tag{7}$$

and

$$\beta_{33} = 0. \tag{8}$$

In eqns (6) and (7) g_1 and g_2 are the central and peripheral CO_2 sensitivities. The central input is

$$U_1(n) = P_{ET,CO_2}\{t(n) - T_1\} - k. \tag{9}$$

The peripheral input is

$$U_2(n) = P_{ET,CO_2}\{t(n) - T_2\} - k \tag{10}$$

and

$$U_3(n) = 0. \tag{11}$$

In eqn (9) T_1 is the central transport delay time and in eqn (10) T_2 is the peripheral transport delay time. In eqns (9) and (10) k is an offset. The noise sequences $V(n)$ and $W(n)$ are independent zero mean white noises with variances q_{11}, q_{22}, q_{33} and r. Finally D is a diagonal matrix with elements d_{11}, d_{22} and d_{33}. The input functions $U_i(n)$ are obtained from the measured $P_{ET,CO_2}(t(n))$, delayed by T_i, using linear interpolation to match the measured ventilation points. We now consider three different models by specifying the matrix D and the measurement matrix C:

I. The measurement noise model M_1,

$$C = [\,1,\,1,\,0];\ d_{11}\ =\ d_{22}\ =\ d_{33}\ =\ 0.$$

II. The process noise model M_2,

$$C = [\,1, 1, 0\,]; \quad d_{11} = d_{22} = 1 \text{ and } d_{33} = 0.$$

III. The external noise model M_3,

$$C = [\,1, 1, 1\,]; \quad d_{11} = d_{22} = 0 \text{ and } d_{33} = 1.$$

PARAMETER ESTIMATION

We estimate the parameters of the three models using a one-step prediction error method[7]. To do this the model equations are written in the innovations form

$$\hat{X}(n+1) = A(n+1)\hat{X}(n) + B(n+1)U(n+1) + K(n)\epsilon(n) \tag{12}$$

and

$$Y(n) = C\hat{X}(n) + ct(n) + \epsilon(n). \tag{13}$$

Here the Kalman gain $K(n)$ is given as

$$K(n) = A(n+1)P(n)C^{\mathsf{T}} \left[CP(n)C^{\mathsf{T}} + 1 \right]^{-1}, \tag{14}$$

where $P(n)$ is obtained as the positive semidefinite solution of the Riccati equation

$$P(n+1) = A(n+1)P(n)A^{\mathsf{T}}(n+1) + DQ/r - K(n)[CP(n)C^{\mathsf{T}} + 1]K^{\mathsf{T}}(n). \tag{15}$$

In eqn (15) Q is an diagonal matrix with elements q_{ii} $(i = 1, 2, 3)$. The matrix $rP(n)$ is the covariance matrix of the state estimate error

$$rP(n) = E\left\{ \left[X(n) - \hat{X}(n) \right] \left[X(n) - \hat{X}(n) \right]^{\mathsf{T}} \right\}, \tag{16}$$

with E the expectation value operator. The innovations $\epsilon(n)$ are a white gaussian noise sequence. The parameter vector Θ_1 for M_1 is

$$\Theta_1 = \{X_{10}, X_{20}, T_1, T_2, k, g_1, g_2, m, b, \alpha_2, c\}^{\mathsf{T}}, \tag{17}$$

for M_2

$$\Theta_2 = \{\Theta_1^{\mathsf{T}}, q_{11}/r, q_{22}/r\}^{\mathsf{T}} \tag{18}$$

and for M_3

$$\Theta_3 = \{\Theta_1^{\mathsf{T}}, q_{33}/r, \alpha_3\}^{\mathsf{T}}. \tag{19}$$

Since it is reasonable to propose that a good model has small prediction errors $\epsilon(n)$, we minimize the cost function

$$J = \sum_{i=0}^{N-1} \epsilon(n)^2. \tag{20}$$

The initial conditions X_{10} and X_{20} at the initial time are taken to be constants and parameters to be estimated. For M_1 the Kalman gain $K(n)$ is identically zero for all n and initial conditions. The initial condition X_{30} in M_3 is assumed to be zero. The initial covariance matrix P_0 for M_2 and M_3 is taken as the identity matrix.

We use a Levenberg-Marquardt algorithm and analytically calculated cost function derivatives. Since the input U_i is obtained by linearly interpolating between measured P_{ET,CO_2} points, the cost function is nonsmooth with regards to the delay parameters T_1

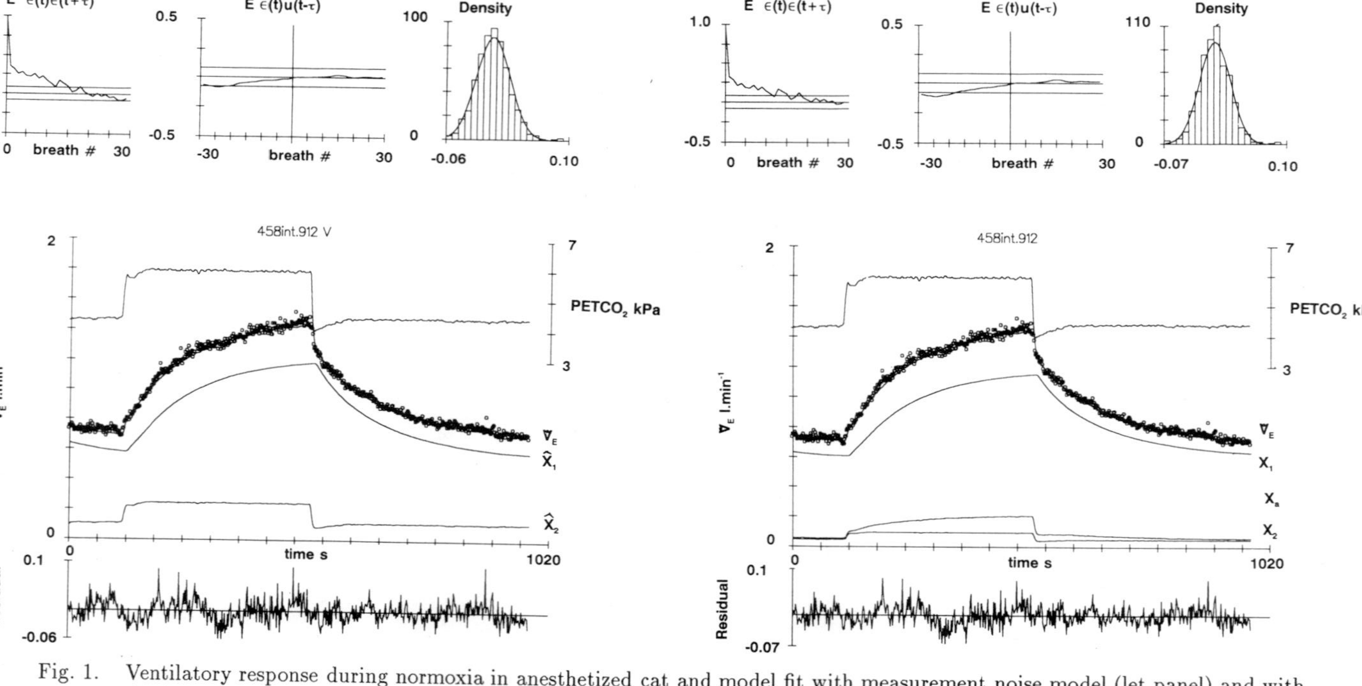

Fig. 1. Ventilatory response during normoxia in anesthetized cat and model fit with measurement noise model (let panel) and with the interaction model (right panel).

Left panel: auto–correlation function of the innovations (top left), cross–correlation function of the innovations and input $PETCO_2$ (top middle) and amplitude distribution of the innovations with estimated Gaussian distribution (top right). Horizontal lines denote 95% confidence limits. Middle figure shows the $PETCO_2$ stimulus, breath–to–breath ventilation (dots). The smooth curve running through the dots is the model fit; it is the sum of the slow central component $\hat{X}_1$, the fast peripheral component $\hat{X}_2$ and the trend (not shown). The bottom figure shows the innovations (residuals).

Right panel: same as left panel but X_1 a central, X_2 a peripheral and X_a an interaction component.

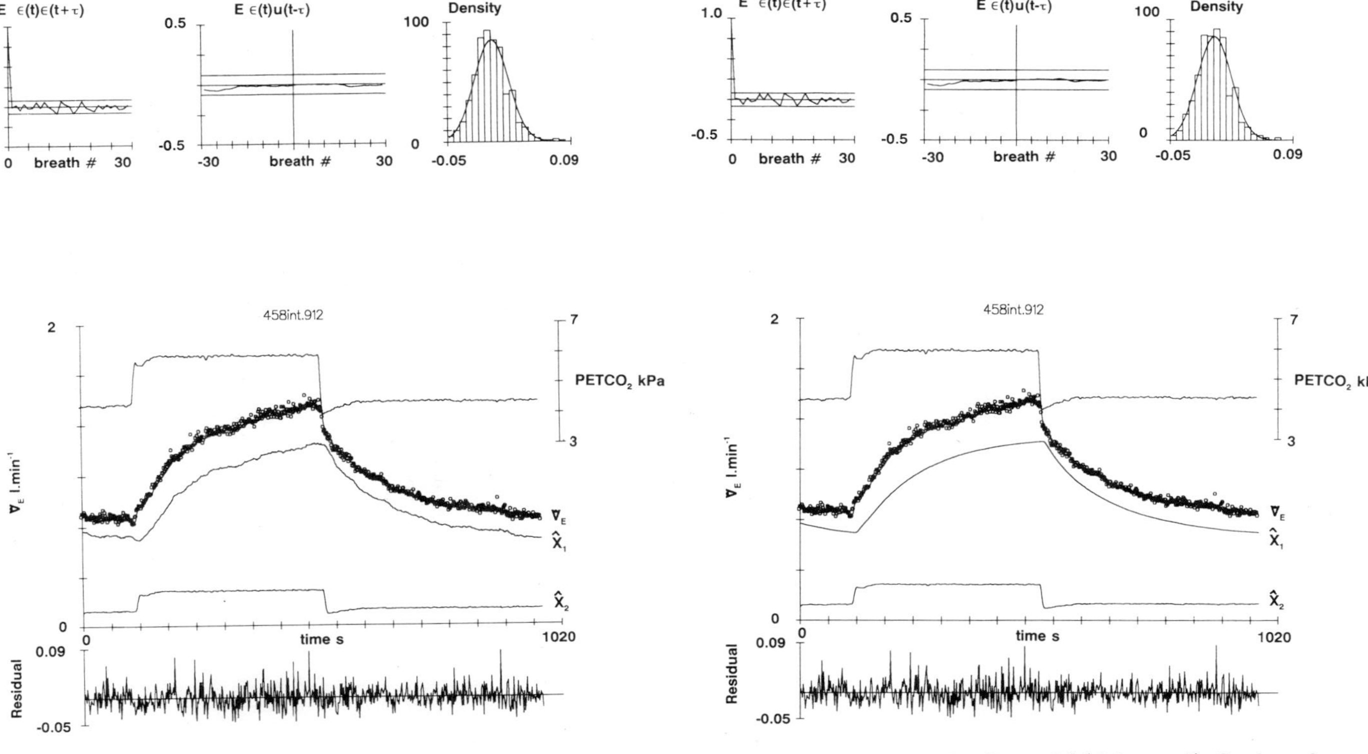

Fig. 2.   Analysis of the ventilatory response with process noise model (left panel) and external noise model (right panel). See legend of Fig. 1 for explanation of the panels. Same experiment as in Fig. 1. Note the process noise superimposed on $\widehat{X}_1$ and $\widehat{X}_2$ and that the auto–correlation functions look white. Component $\widehat{X}_3$, not drawn in right panel.

and T_2. To avoid numerical difficulties in an algorithm which needs continuous first derivatives of the cost function[8], the time delays T_1 and T_2 are not estimated on the same footing as the other parameters. To obtain optimal delay times an "exhaustive" grid search is applied. For all combinations of time delays with the constraint $T_1 \geq T_2$, the cost function J is minimised with respect to the other parameters untill an "overall" minimum is obtained. This method is similar to using concentrated likelihood functions[8].

MODEL VALIDATION

Using a square wave input P_{ET,CO_2} against a constant background of P_{ET,O_2} the central and peripheral effects of CO_2 on ventilation have been studied in human beings[6] and anesthetised cats[1]. By combining a technique of artificial perfusion of the brain stem and the DEF technique it has been shown that the measurement noise model adequately describes the ventilatory response of the peripheral and central chemoreflexes to square waves of P_{ET,CO_2}[1,10,11]. In general the residuals of M_1 are clearly non-white in awake human beings and anesthetised cats (see Figs. 1 and 3), although the cross correlation between residuals and input function is often quite small suggesting that the model has extracted all the information from the measured ventilation (see Figs. 1 and 3). This is somewhat strengthened by ensemble averaging the difference between the measured ventilation and the ventilation from the noise free model using the actual input P_{ET,CO_2}[9,10,11]. These ensemble averages do not show systematic deviations indicative of a model misspecification. Our calculation of the asymptotic covariance matrix of the parameters is based, among other things, on residuals which are independent. Fitting simulated data corrupted with coloured noise with the measurement noise model suggests that the variances of the parameters are underestimated several times. We remark that the variances of the parameters are in general underestimated, since in the estimation procedure they are "conditional" on the values of the time delays. Although an estimate of the standard deviations of the time delays T_1 and T_2 could be obtained from the results of the grid search, we have not attempted this. The process noise model and the external noise model both lead to "white" innovations and small correlations between inputs and innovations (see Figs. 2 and 4). The values of the parameters common to the three models are quite similar. However, the estimated variances are usually smaller with the measurement noise model compared to the other models. In the external noise model, the noise is parametrized independent of the systematic (deterministic) part of the model in contrast to the process noise model. Under suitable conditions the parameters of the systematic part of the model are estimated correctly even if the noise is modelled inadequately[7]. It would be interesting to study the above mentioned aspects using simulated data from the additive two-compartment model. Also judging parameter accuracy by graphing sensitivities of both the data and the analytic model[6] and using "exact" confidence regions[8] would be quite interesting. Although it appears that the measurement noise model is not quite satisfactory, it is still unclear whether the process noise or external noise model is to be preferred from a physiological point of view, since the peripheral time constant τ_2 ($1/\alpha_2$) is often quite close to the time constant $1/\alpha_3$ in the external noise model. For a discussion of this point we refer to Ward et al.[12]. To give an impression of the values of the model parameters together with their standard deviations in anesthetized cats and awake human beings we collected the physiological most interesting ones in Table 1.

APPLICATIONS

The DEF technique has been used in normal subjects and subjects who had theur carotid bodies resected. It was found, among other things, that the time constant of the ventilatory on-transient was markedly longer than the one of the ventilatory off-transient in both groups[6]. This could be caused by changes in cerebral blood flow together with neuronal dynamics[1]. However, in a recent study Dahan et al. reported no significant difference between τ_{on} and τ_{off} in awake human subjects during normoxia[9]. In anesthetized cats the time constant of the off-transient was found to be significantly larger than the on-transient[1]. In awake human

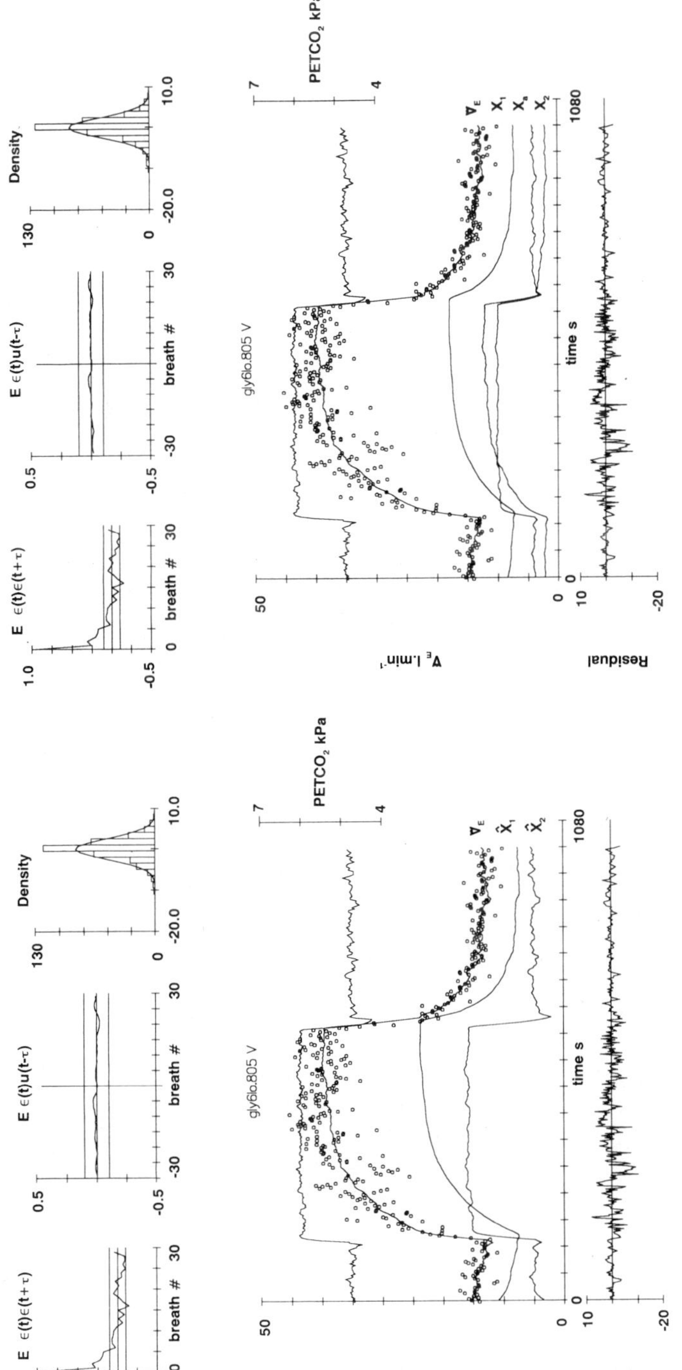

Fig. 3. Ventilatory response during mild hypoxia in awake human subject analyzed with measurement noise model (left panel) and interaction model (right panel). For further explanation see Fig. 1.

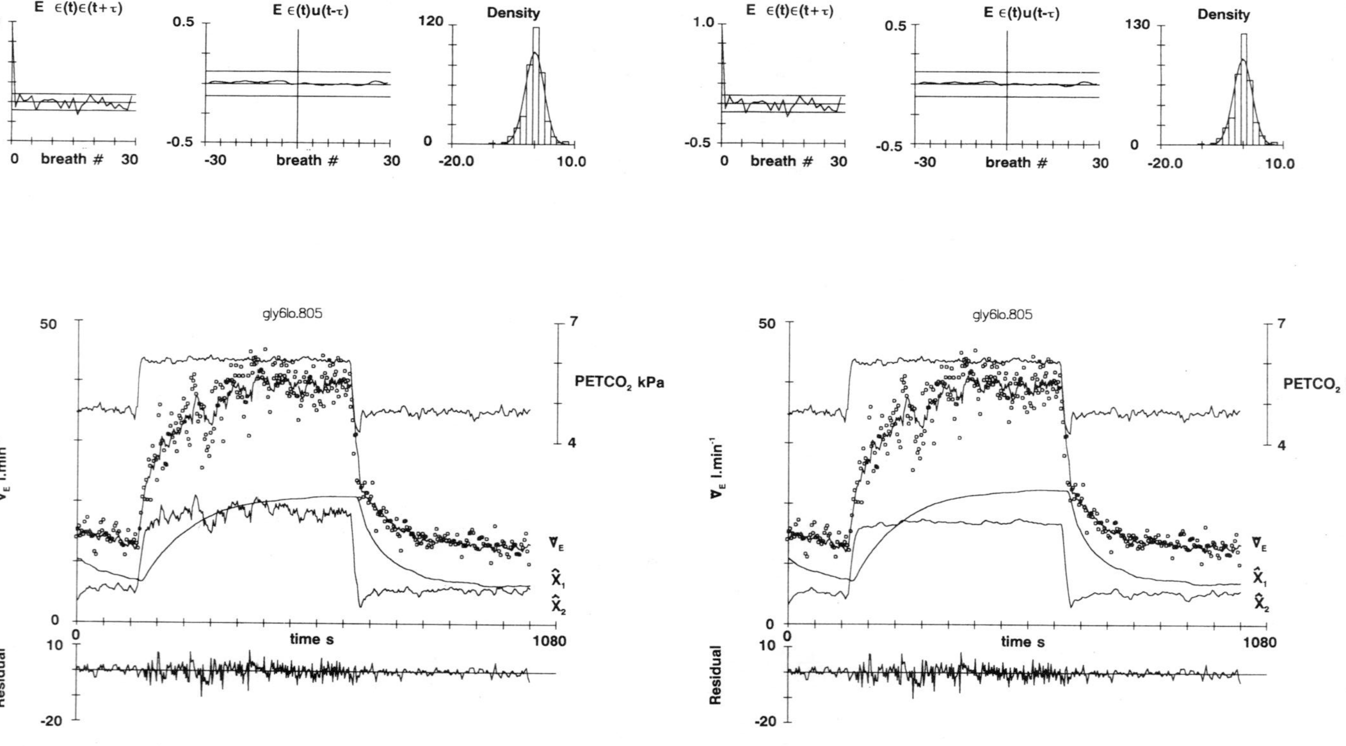

Fig. 4. Analysis of the ventilatory response with process noise model (left panel) and external noise model (right panel). Same experiment as in Fig. 3. For further explanation see Fig. 1.

Table 1. Estimated parameters $\pm$ standard deviation of Model 3. k in kPa, g_1 and g_2 in l min^{-1} kPa^{-1}, $\tau_{on}, \tau_{off}, \tau_2, T_1, T_2$ in s and c in l min^{-2}. Cat data obtained from analysis shown in right panel of Fig. 2, human data from right panel of Fig. 4.

	cat data			human data		
k	3.06	$\pm$	0.10	4.20	$\pm$	0.10
g_1	0.42	$\pm$	0.01	12.3	$\pm$	1.4
g_2	0.08	$\pm$	0.01	9.3	$\pm$	1.7
τ_{on}	130	$\pm$	11	95	$\pm$	25
τ_{off}	147	$\pm$	10	68	$\pm$	16
τ_2	1.6	$\pm$	0.55	8.6	$\pm$	3.4
T_1	11.0			15.0		
T_2	4.0			3.0		
c	1.16	$\pm$	2.14	37.7	$\pm$	11.1

subjects intravenous dopamine caused a reduction of the peripheral gain while there was no change in the central gain[13]. In contrast, there was only a change in offset in anesthetized cats[14]. Unfortunately the DEF technique is unable to distinguish between a change in offset due to the peripheral or central chemoreflex. If we introduce two different offsets, one for the peripheral chemoreflex and one for the central chemoreflex, it can be readily shown that they are not identifiable. This is the reason that in the two-compartment model the offsets of the peripheral and central chemoreflexes are chosen equal. To decide whether a change in offset is of central or peripheral origin additional information is required. In anesthetized cats intravenously administered domperidone only decreases the offset k, while haloperidol also diminishes the central and peripheral gain[15]. Almitrine decreases the threshold and augments the peripheral gain g_2, but leaves the central gain g_1 unaltered[16].

These drug studies illustrate that the DEF technique is a powerful tool in separating central end peripheral effects on ventilation.

INTERACTIONS

A problem which solution has been elusive up till now, is that of interaction between peripheral and central chemoreflexes. Recently Robbins[17] reported evidence for an interaction between the contributions to ventilation from the central and peripheral chemoreceptors in awake man and proposed a multiplicative model for the respiratory controller. Following his suggestion we generalize the additive two-compartment model by replacing eqn (3) by

$$Y(n) = X_1(n) + X_2(n) + X_a(n) + ct(n) + W(n), \tag{21}$$

in which X_a is the interaction component. It is given by

$$X_a = g_m X_1(n) X_2(n), \tag{22}$$

where g_m is the strength of the interaction. We assume that there is only measurement noise on the output. This multiplicative model has the property that the apparent peripheral CO_2 sensitivity (and also the apparent central CO_2 sensitivity) will be different for a step increase in P_{ET,CO_2} than for a step decrease in P_{ET,CO_2}. This asymmetry in CO_2 sensitivity may be characterized by the asymmetry parameter

$$g_a = g_m g_1 g_2. \tag{23}$$

In Figs. 1 and 3 we give examples of experiments analyzed with this interaction model. In these examples the asymmetry parameter, which indicates a positive interaction, is significantly different from zero. However, in a preliminary analysis of the ventilatory response following square wave changes in P_{ET,CO_2} in normoxia we could not find clear evidence for such an interaction in awake human subjects and anesthetized cats. It may well be that the data sets obtained from the experiments using square wave P_{ET,CO_2} challenges are not informative enough to unearth such an interaction. The question arises how to develop optimal experimental designs for discriminating between several competing models. Little work has been done in this direction in respiratory control (cf. ref. 3).

CONCLUDING REMARKS

Since the introduction of the dynamic end-tidal forcing technique by Swanson and Bellville[4], its application has extended our knowledge of the ventilatory controller and of the actions of various drugs. The additive two-compartment model has been succesful in extracting useful physiological information. It should be emphasized, however, that most experiments have been performed with square wave changes in P_{ET,CO_2}. Testing the additive two-compartment model with sinusoidal changes in end-tidal P_{CO_2}, Robbins (ref. 18 and references cited therein) found that the phase shifts of the measured frequency responses is much smaller than those predicted by the model. Also the application of a step change in P_{ET,CO_2} followed by a linear rate of rise in man leads to results which seem to be inconsistent with the additive two-compartment model[19]. The issue is whether the time constant deduced from experiments with step-ramp inputs is consistent with those obtained from square wave inputs. The problem how the present respiratory two-compartment model has to be modified to resolve these difficulties is an important area of future work.

REFERENCES

1. J. DeGoede, A. Berkenbosch, D.S. Ward, J.W. Bellville and C.N. Olievier, Comparison of chemoreflex gains obtained with two different methods in cats, *J. Appl. Physiol.* 59: 170-179 (1985).

2. M.C.K. Khoo and S.M. Yamashiro, Models of control of breathing, *in*: "Respiratory Physiology: an analytical approach, vol 40, Lung Biology in Health and Disease," H.K. Chang and M. Paiva, eds., Marcel Dekker, New York (1989).

3. J.W. Bellville, D.S. Ward and D. Wiberg, Respiratory System: Modelling and identification, *in*: "Systems and Control Encyclopedia: Theory, Technology, Applications," H.G. Singh, ed., Pergamon Press, Oxford (1988).

4. G.D. Swanson and J.W. Bellville, Step changes in end-tidal CO_2: Methods and implications, *J. Appl. Physiol.* 39: 377-385 (1975).

5. D.M. Wiberg, J.W. Bellville, O. Brovko, R. Maine and T.C. Tai, Modelling and parameter identification of the human respiratory system, *IEEE Trans. Auto. Cont.*: 24, 716-720 (1979).

6. J.W. Bellville, B.J. Whipp, R.D. Kaufman, G.D. Swanson, K.A. Aqleh and D.M. Wiberg, Central and peripheral chemoreflex loop gain in normal and carotid body-resected subjects, *J. Appl. Physiol.* 48: 843-853 (1979).

7. L. Ljung, "System Identification: Theory for the user," Prentice-Hall, Inc., Englewood Cliffs, New Jersey (1987).

8. G.A.F. Seber and C.J. Wild, "Nonlinear regression," John Wiley and sons, New York (1989).

9. A. Dahan, I.C.W. Olievier, A. Berkenbosch and J. DeGoede, Modelling the dynamics of the ventilatory response to carbon dioxide in healthy human subjects during normoxia, *in*: "Respiratory Control: A Modelling Perspective," G.D. Swanson and F.S. Grodins, eds., Plenum Publishing Co., New York, in press.

10. A. Berkenbosch, J. DeGoede, D.S. Ward, C.N. Olievier and J. VanHartevelt, Dynamic response of the peripheral chemoreflex loop to changes in end-tidal CO_2, *J. Appl. Physiol.* 64: 1779-1785 (1988).

11. A. Berkenbosch, D.S. Ward, C.N. Olievier, J. DeGoede and J. Hartevelt, Dynamics of ventilatory response to step changes in P_{CO_2} of blood perfusing the brain stem, *J. Appl. Physiol.* 66: 2168-2173 (1989).

12. D.S. Ward, J. DeGoede, D.M. Wiberg, A. Berkenbosch and J.W. Bellville, Analysis of a ventilatory noise model in man and cats, *in*: "Modelling and the control of breathing", B.J. Whipp and D.M. Wiberg, eds., Elsevier Publishing Co., Amsterdam (1983).

13. D.S. Ward and J.W. Bellville, Effect of intravenous dopamine on hypercapnic ventilatory response in humans, *J. Appl. Physiol.* 55: 1418-1425 (1983).

14. A. Berkenbosch, J. DeGoede, C.N. Olievier and D.S. Ward, Effect of exogenous dopamine on the hypercapnic ventilatory response in cats during normoxia, *Pfluegers Archiv* 407: 504-509 (1986).

15. A. Berkenbosch, C.N. Olievier and J. DeGoede, Effects of dopamine antagonists haloperidoland domperidone on the normoxic ventilatory response to CO_2 in cats, *Pfluegers Archiv* 411: 278-282 (1988).

16. C.N. Olievier, A. Berkenbosch, J. DeGoede and E.W. Kruyt, Almitrine bismesylate and the central and peripheral ventilatory response to CO_2, *J. Appl. Physiol.* 63: 66-74 (1987).

17. P.A. Robbins, Evidence for interaction between the contribution to ventilation from the central and peripheral chemoreceptors in man, *J. Physiol.* 401: 503-518 (1988).

18. P.A. Robbins, The ventilatory response of the human respiratory system to sine waves of alveolar carbon dioxide and hypoxia, *J. Physiol.* 350: 461-474 (1984).

19. A. Berkenbosch, J.G. Bovill, A. Dahan, J. DeGoede and I.C.W. Olievier, The ventilatory CO_2 sensitivities from Read's rebreathing method and the steady-state method are not equal in man, *J. Physiol.* 411: 367-377 (1989).

DYNAMIC MODELS AND PARAMETER ESTIMATION:

THE HYPOXIC VENTILATORY RESPONSE

Denham S. Ward

Departments of Anesthesiology and Electrical Engineering
University of California, Los Angeles
Los Angeles, CA 90024-1778

INTRODUCTION

Mathematical models of the control of breathing during dynamic changes in various stimuli (*e.g.*, CO_2, O_2 and exercise) have been extremely useful [2] in understanding how ventilation is adjusted. In the physical sciences models can often be derived from fundamental physical relationships (*e.g.*, the equations of motion for satellite orbits) but in physiology there are few such relationships to guide the model builder. The building of a model can be divided into two stages. First the structure of the model must be determined. The structure consists of the mathematical equations and any parameters that are not estimated from an individual data set (assumed values and known constants). Secondly, experiments and parameter estimation techniques are used to determine the values of certain parameters from individual experiments. The validation of the model using different data sets and experimental conditions can then be done. Frequently this validation process will suggest modifications to the model and the process starts over. This paper will discuss how some of the assumptions that must go into devising a model for the hypoxic ventilatory response determine the characteristics of such a model. Ultimately a mathematical model is useful to summarize and predict the response and how it is changed by pathological, physiological or pharmacological interventions.

QUALITATIVE CHARACTERISTICS

The qualitative characteristics of the hypoxic ventilatory response will determine what types of mathematical models can be considered. There are two prominent characteristics of the hypoxic response that must be incorporated into the model. The first is the well known nonlinear steady state relationship between oxygen tension and ventilation. Until the oxygen tension falls below approximately 60 – 70 mm Hg there is little increase in ventilation. However, lower tensions produce a marked increase in ventilation. Various models for this nonlinear response have been proposed: hyperbolic, exponential and linear in hemoglobin saturation [15]. Each of these models can describe the data adequately and each has 2 or 3 free parameters to estimate from the data.

The second characteristic that is important to model is the biphasic nature of the ventilatory response when there is a step decrease in oxygen. Since the carbon dioxide tension will affect the ventilation, all experiments must keep the end-tidal CO_2 (P_{ETCO_2}) constant by appropriate adjustment of the inspired carbon dioxide. The ventilatory response to an isocapnic step decrease in end-tidal O_2 (P_{ETO_2}) shows a rapid increase to a plateau that maintains a constant ventilation for 5 to 10 minutes. Thereafter there is a slow decline to a final value that is usually intermediate between the peak hypoxic ventilation and the normoxic ventilation. The response to the step out of hypoxia is similar but there are apparent

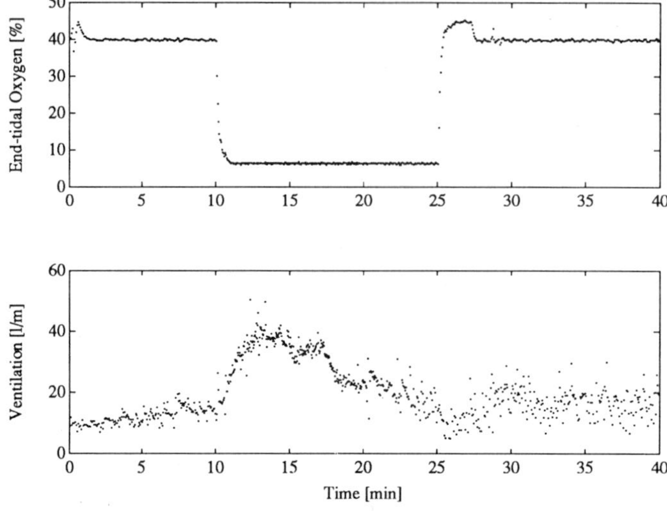

Figure 1. *Typical breath-by-breath ventilatory response to a step into and out of hypoxia, starting and finishing in hyperoxia. Each point is a single breath.*

asymmetries in the response when comparing the step into hypoxia with the step out of hypoxia. Figure 1 shows a typical ventilatory response to a step into hypoxia followed by a step out of hypoxia. This response shows the marked overshoot in the ventilation going into hypoxia while there is very little undershoot on the way out of hypoxia. While there is considerable individual variation in the hypoxic response, with some subjects showing very little depression and others not even showing much stimulation, the essential features are shown in Fig. 1. At least in adult subjects, the depression does not ordinarily overcome the initial stimulation and the plateau ventilation remains above the pre-hypoxic level.

Besides these primary characteristics there are other factors that may require incorporation into the model at some stage in model development. These include a lasting effect upon ventilation after the return to normoxia and a possible correlation between the magnitude of the initial increase and the subsequent decline [6, 7, 8, 9]. However, when constructing a model it is important to initially select the most salient features. Trying to be too inclusive results in a model that does not lend itself to parameter identification from individual experiments. More complex simulation models are useful for summarizing known data and can be used as reference models with which to generate data for testing the parameter identification schemes based on simpler models.

MODELLING THE NONLINEARITY

While the form of the nonlinear relationship between the P_{ETO_2} and ventilation has received considerable attention, there is no consensus on the correct choice [15]. In actuality, given the variability in ventilation (noise), the issue cannot be resolved on model fit considerations alone. In this discussion an exponential relationship will be assumed primarily for mathematical convenience. As will be discussed below, the direct estimation of the nonlinear parameters of the steady-state relationship is difficult from dynamic response data sets. No matter what the shape of the steady-state relationship, it may be necessary to assign parameter values that have been determined from other considerations. These assigned values, not estimated from the dynamic data, are part of the structural model and errors in the value

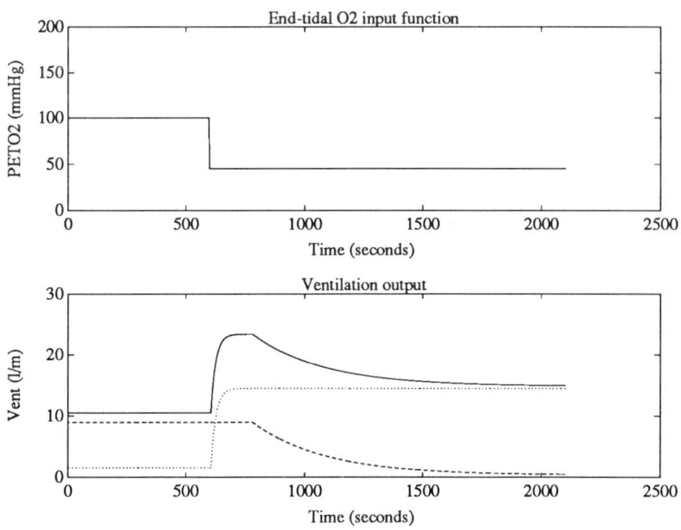

Figure 2. *Response of the model to a step decrease in oxygen. The total response is shown by the solid line, the peripheral component, X_s, by the dotted line and the central component, $Y_o - X_d$, by the dashed line. The model parameters are:* $g_s = 90\ l \cdot min^{-1}$, $g_d = 60\ l \cdot min^{-1}$, $\tau_s = 20.0\ s$, $\tau_d = 300.\ s$, $\Delta_s = 9.0\ s$, $\Delta_d = 183\ s$, $Y_o = 10\ l \cdot min^{-1}$.

will cause distortions in the estimation of the other parameters.

Two issues become apparent when this steady-state model is extended into a dynamic model. The first is that while the steady-state relationship for the ventilatory response to hypoxia is well described, this response is made up of both the ventilation increasing and the ventilation decreasing effects of hypoxia. It is not known if the ventilatory increase and decline separately have the same steady-state response shape in humans. Animals studies indicate that most likely the shape is qualitatively the same but whether or not the parameters are the same is unknown [1, 3].

The second issue is the relationship between the nonlinearity and the dynamics. That is, does the nonlinearity occur before or after the dynamics? The answer to this question has both modelling and physiological considerations. For example, if the oxygen tension (arterial P_{O_2} as estimated by the measurement of P_{ETO_2}) undergoes the nonlinear transformation at the chemoreceptor, then the occurrence of the dynamics prior to the transformation would imply that the dynamics are caused by the transport characteristics for oxygen between the lungs and the receptor. If the dynamics are assumed to occur after the transformation at the chemoreceptor, then this would imply that the dynamics are caused by neural processing [10]. Simulations can be used to design experiments to help distinguish between these model structures.

THE DYNAMIC MODEL AND PARAMETER ESTIMATION

A straightforward two-component dynamic model has been used previously to model the hypoxic response in cats with some success [3, 4]. This model can be applied to human data with the appropriate changes in the parameter values.

All variables in this model are on a breath-by-breath basis, and are assumed constant over that breath. T_N is the length of the Nth breath. Since the steady-state response to

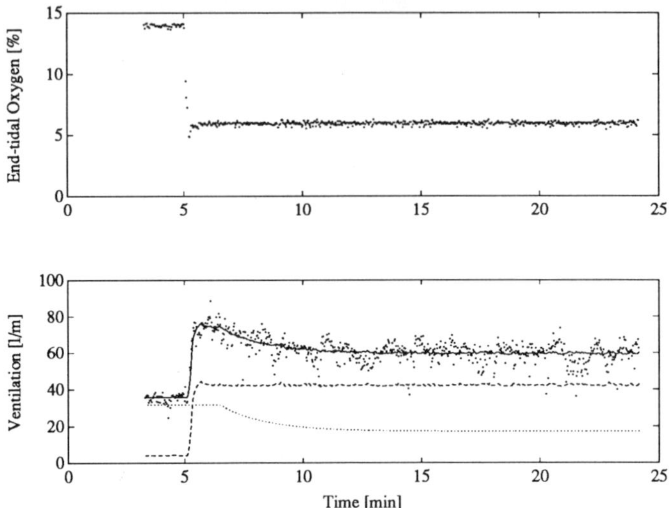

Figure 3. *Model fit to ventilatory data from a step into hypoxia. The points are the breath-by-breath measurements. The solid line through the breath-by-breath ventilation is the least squares fit of the model. The two components of the model are shown by the dotted and the dashed lines. The model parameters are: $g_s = 241\ l \cdot min^{-1}$, $g_d = 92\ l \cdot min^{-1}$, $\tau_s = 8.9\ s$, $\tau_d = 121\ s$, $\Delta_s = 3.7\ s$, $\Delta_d = 84\ s$, $Y_o = 33\ l \cdot min^{-1}$.*

hypoxia is a nonlinear function of the P_{ETO_2}, this transformation must be made to the measurements. As discussed in the previous sections, there is no general agreement as to the exact form for this nonlinearity, but animal studies [1] indicate that an exponential form for both the stimulation and the depressive aspects of the ventilatory response to hypoxia is adequate. Equation (1) gives this relationship with "D" as the parameter. Since for simple single step input, it will not be possible to accurately estimate D, we must pick a value. It has been well described that the ventilatory response to hypoxia is linear when plotted against oxygen saturation. In the range above 40 mm Hg, the standard saturation curve is quite well fit by the simple exponential model. This results in a value of D equal to 0.041 mm Hg^{-1}. The shape of the central decline is not known in humans but animal data would indicate that the decline is related to oxygen content and not partial pressure. If this is the case then the assumption of an exponential shape for the central component may be reasonable. Since no information is available, it is assumed that the shape and the parameter value D is the same for both the stimulating and the depressive inputs. Further work in humans is needed to clarify these aspects of the steady-state relationship. The input to the model is assumed to be U(N) and D is taken to be 0.041 mm Hg^{-1} in all subsequent model analyses and simulations.

$$U(N) = \exp[-D \cdot P_{ETO_2}(N)] \tag{1}$$

The state X_s represents the ventilatory stimulating effects of hypoxia and X_d represents the ventilatory decreasing or depressive effects of hypoxia.

$$X_s(N + 1) = \alpha_s(N)\, X_s(N) + g_s[1 - \alpha_s(N)]\, U(N - \Delta_s) \tag{2}$$

$$X_d(N+1) = \alpha_d(N)\, X_d(N) + g_d[1 - \alpha_d(N)]\, U(N - \Delta_d) \qquad (3)$$

The breath-by-breath transition coefficients are determined by the constants τ_s and τ_d and the length of each breath.

$$\alpha_s(N) = \exp[-T_N/\tau_s] \qquad (4)$$

$$\alpha_d(N) = \exp[-T_N/\tau_d] \qquad (5)$$

The gains for each state equation are g_s and g_d. Note that the exponential transformation of the end-tidal P_{O_2} (Eq. 1) makes the units of g_s and g_d simply $1 \cdot \text{min}^{-1}$. The values of the gains thus represent the maximum steady-state change in the state variable if the P_{ETO_2} could be made zero.

The total ventilation on breath N, $Y(N)$, is represented as the sum of these state variables plus a bias term, Y_o, that is oxygen level independent. This term is required since both X_s and X_d become small at high oxygen levels.

$$Y(N) = X_s(N) - X_d(N) + Y_o \qquad (6)$$

These equations describe an appropriate model representing some of the physiological processes in a way that the parameter values have some meaning. This is as opposed to a "Black Box" model that describes the input-output relationship, but the parameters of the model do not have any physiological interpretation.

Figure 2 shows the response of this model to a step decrease in hypoxia. In this and in the subsequent figures of the model response the two components are plotted as X_s and $Y_o - X_d$. To simplify calculations the length of each breath T_N is taken as equal to 3 seconds independent of the breath number. At least qualitatively, the response of the model matches the response seen in Fig. 1.

A model that incorporates the breath-by-breath variation in the ventilation has not been excluded in this model formulation. In the state space model given in equations (2) to (6), the noise can be incorporated as white noise additive to the output, additive to the state variable equations, or as a separate correlated noise process added to the output [5, 13].

The residual function is the difference between the measured ventilation and the ventilation predicted by the model with the same input function. Naturally, if the model and reality were identical the residual would be zero throughout the experiment. Two factors contribute to the residual being not identically zero. Any differences between reality and the mathematical model (modelling error) that occurs with the particular input function used will show up as deviations from zero. In addition, there will be breath-to-breath variation in ventilation (noise) that will cause random variations in the residual. Noise models can model the statistics of these random variations, but if they are truly random then the exact realization for the particular experiment will not be duplicated.

Often because of the noisy ventilation data, it is difficult to detect modelling error in the residual function. Although the effects of the model structural error show up readily when there is no noise in the model, the amount of error is small enough to be easily disguised by the normal breath-to-breath variation in ventilation. With simulations it is easy to study in isolation the effects of a modelling error. The data can be generated without any noise and then the effects of estimating the parameters based on an estimator that assumes the wrong model can be studied. In this situation, incorrect parameters will be estimated and

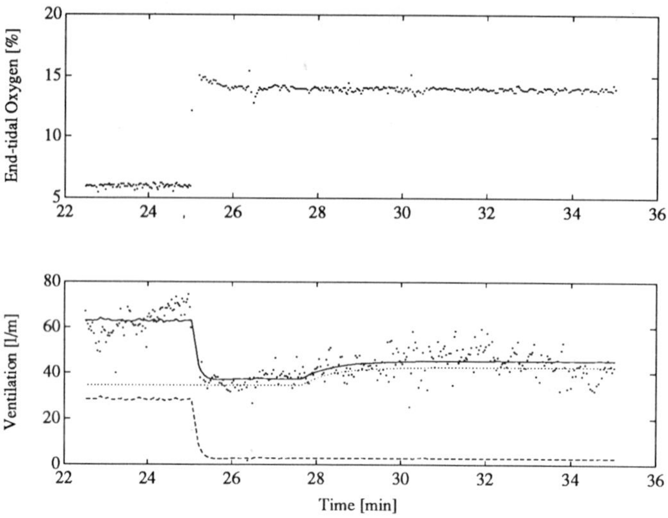

Figure 4. *Model fit to ventilatory data from a step out of hypoxia. The points are the breath-by-breath measurements. The solid line through the breath-by-breath ventilation is the least squares fit of the model. The two components of the model are shown by the dotted and the dashed lines. The model parameters are:* $g_s = 162 \, l \cdot min^{-1}$, $g_d = 51 \, l \cdot min^{-1}$, $\tau_s = 7.1 \, s$, $\tau_d = 44 \, s$, $\Delta_s = 3.0 \, s$, $\Delta_d = 158 \, s$, $Y_o = 44 \, l \cdot min^{-1}$.

the residual will show a systematic pattern [14]. The parameter estimator for this model is based on the one-step predictor technique (see [11, pages 169-197] for a full discussion of this predictor and its relationship to maximum likelihood techniques) and the parameters are determined by minimizing the sum of the squares of the difference between the model prediction of the ventilation and the actual measured ventilation. The actual numerical minimization uses a Levenberg-Marquardt procedure [11].

FITTING REAL DATA

Once the structure of the model has been determined and the parameter estimation scheme developed, it is then necessary to use the model on a variety of data sets for validation, possible modifications and hopefully to gain physiological insight into the mechanisms of the hypoxic ventilatory response.

This model has been used to compare the asymmetry between the transition into hypoxia and the transition out of hypoxia, the effects of different P_{ETCO_2} levels and the effect of exercise. Generally step inputs have been used, but occasionally when a step has not been precisely achieved the model still accomplishes a good fit to the data. Figures 3 and 4 give typical fits to steps into and steps out of hypoxia.

While these are apparently reasonable fits, when the two fits are examined together a difficulty becomes apparent. While the step into and the step out of hypoxia have been treated separately, obviously in reality one follows the other. The curve fits shown in Fig. 4 is the continuation of the data in Fig. 3. Since the fits have been done separately there is no restriction that the final conditions of the two components in Fig. 3 are equal to the initial conditions in Fig. 4. Examining the two figures reveals that in fact there is a discontinuity in the two components if the two fits are rejoined into one data set. We also see that the

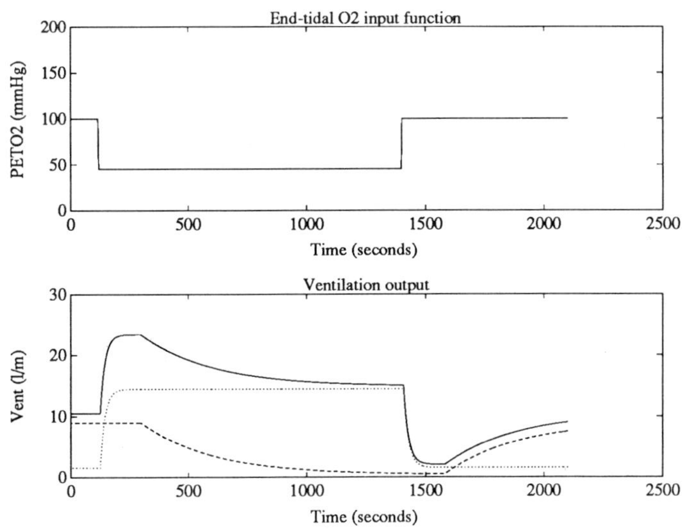

Figure 5. *Model output for a symmetric input. Note that the ventilation is also symmetric and results in a marked, prolonged undershoot with the transition out of hypoxia. The model parameters are given in Fig. 2. Contrast this with the measured response shown in Fig. 1*

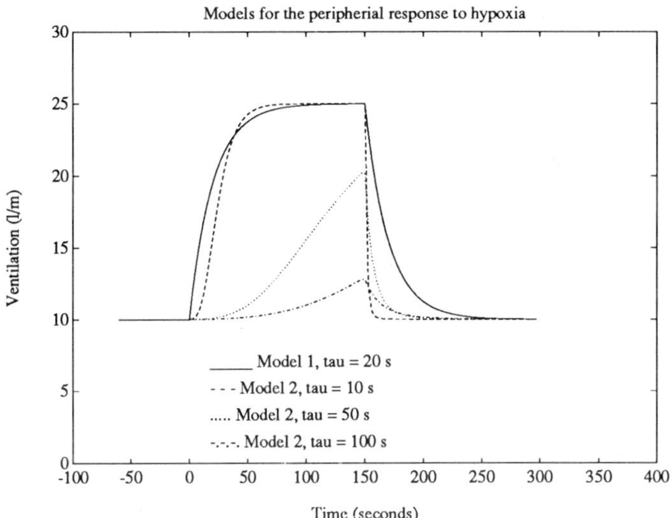

Figure 6. *Comparison of the peripheral response of Model 1 with Model 2 for different time constants in Model 2. No time delay is included in either model. The P_{ETO_2} input (not shown) is a step into hypoxia at time 0 followed by a step out at 150 s. Note that for the step into hypoxia a 10 s time constant for Model 2 closely matches a 20 s time constant for Model 1. However for the step out of hypoxia even with a 100 s time constant for Model 2 results in a faster ventilatory response than the 10 s time constant for Model 1.*

parameter values are different. Both of these observations come from the fact that while our model predicts a symmetric response when the input is symmetric, the data is not symmetric (Fig. 2 shows a different example data set). When several data sets are examined we find that the average parameters found for the step into hypoxia do not equal the average parameters for the step out of hypoxia [12].

By plotting the simulation for a step into hypoxia followed by a step out of hypoxia, the effects of the symmetry in the ventilation are apparent (Fig. 5). At this point the validity and usefulness of the model must be assessed. Although a reasonable fit has been obtained for both steps into and steps out of hypoxia, it will be necessary to modify the model in order to obtain the asymmetric response that is typically seen in human subjects. An asymmetry in the central time constant of the CO_2 response was modelled by making the central time constant dependant on the P_{ETCO_2}[2]. However in the case of the hypoxic response the asymmetry is more severe and we have little in the known physiology to guide us as to what are potential alternate models. Model building is useful because of its ability to investigate alternate model structures and, by fitting multiple data sets, to discover a form that can potentially explain the measurements. In this situation the modelling precedes the physiological knowledge and may allow some insight into the physiology and guide the design of further experiments.

MODIFICATIONS TO THE MODEL

Since a nonlinear model is already being considered, there is considerable latitude for modifications to the model to produce an asymmetric response. When a linear model is used the order or sequence of the blocks in a pathway does not effect the input-output response. This is not true in a nonlinear model. The presence of nonlinearities in the system results in profound changes in the input-output response when the sequence of blocks in the model are changed. The model we have discussed (which we will call Model 1 for convenience) assumes that the P_{ETO_2} is transformed by the nonlinear function prior to any time dynamics (see Eq. 1). Although in Model 1 the time delay was placed after the nonlinearity (Eqs. 1, 2 and 3), the location of the time delay does not effect the response. However, if the sequence of the dynamics and the nonlinearity are reversed such that the P_{ETO_2} goes through the time dynamics prior to the nonlinear transformation into the ventilation, then although the steady-state relationship is unaltered, there are major changes to the dynamic response. The model equations which for convenience we will call Model 2, then become:

$$U(N) = P_{ETO_2}(N) \tag{7}$$

$$X'_s(N + 1) = \alpha_s(N)\, X'_s(N) + [1 - \alpha_s(N)]\, U(N - \Delta_s) \tag{8}$$

$$X'_d(N + 1) = \alpha_d(N)\, X'_d(N) + [1 - \alpha_d(N)]\, U(N - \Delta_d) \tag{9}$$

$$X_s(N) = g_s \cdot \exp[-D \cdot X'_s(N)] \tag{10}$$

$$X_d(N) = g_d \cdot \exp[-D \cdot X'_d(N)] \tag{11}$$

$$Y(N) = X_s(N) - X_d(N) + Y_o \tag{12}$$

The time constants are still given by eqs. 4 and 5.

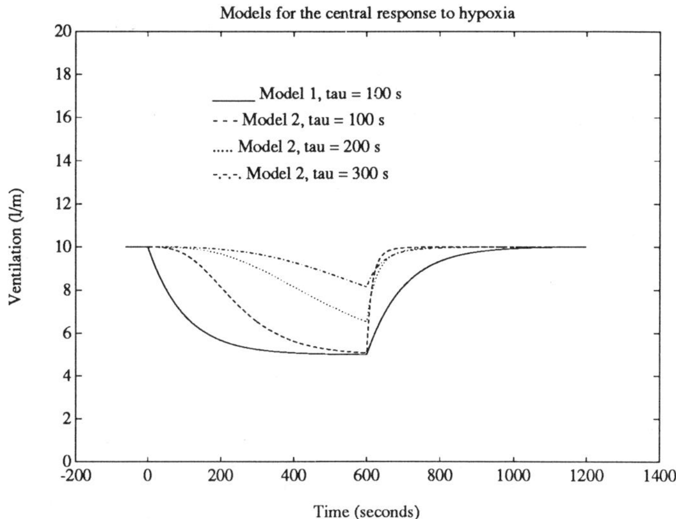

Figure 7. *Comparison of the central response of Model 1 with Model 2 for different time constants in Model 2. No time delay is included in either model. The P_{ETO_2} input (not shown) is a step into hypoxia at time 0 followed by a step out at 150 s. As in Fig. 6 Model 2 results in an asymmetric response with the time course of the ventilation during the step out of hypoxia being much faster than the time course for the step into hypoxia. Note also in Model 2 the shape of the ventilatory response for the transition into hypoxia has an apparent latency before the decline even for the 100 s time constant even though Model 2 has no pure time delay.*

In order to get an idea of how this model responds, the stimulating and the depressive components are examined separately. Figure 6 shows the peripheral response for Model 1 using the same parameters as in Figs. 2 and 5. As a comparison the responses of Model 2 with different time constants (τ_s) are shown. The marked asymmetry in the response of Model 2 for all time constants is readily apparent.

The cause of this asymmetry is the nonlinearity given in Eq. 10. For Model 2, even though the input is a simple step, the occurrence of the dynamic response (Eq. 8) prior to the nonlinearity exposes the full effect of the nonlinearity. During the initial period of the step into hypoxia X_s' changes on the relatively flat portion of the nonlinearity resulting in little change in X_s. When X_s' reaches the steeper portion of the nonlinearity (approximately $3 \cdot D^{-1}$ or 75 mm Hg for our model) the changes in X_s take place faster. For the step out of hypoxia the process is reversed and the faster response comes first. Although data in humans is not available to fit the peripheral response in isolation, Model 2 does not seem appropriate. Certainly, to achieve a reasonable response to a step into hypoxia a time constant would have to be selected that results in a much too rapid response for the step out of hypoxia.

However, the response of Model 2 for the central depressive response is more reasonable. Figure 7 shows a comparison of the response of Model 2 for several values of the time constant to Model 1 with a τ_d of 100 s. The plot is shown for $Y_o - X_d$. For this comparison no time delays are included. It is immediately obvious that even though there is no time delay Model 2 shows an apparently latency before the decline starts. This shape is closer to the observed response which Model 1 had modelled with a long pure time delay. Because of the nonlinearity, the latency will appear longer for steps into hypoxia starting from hyperoxia than

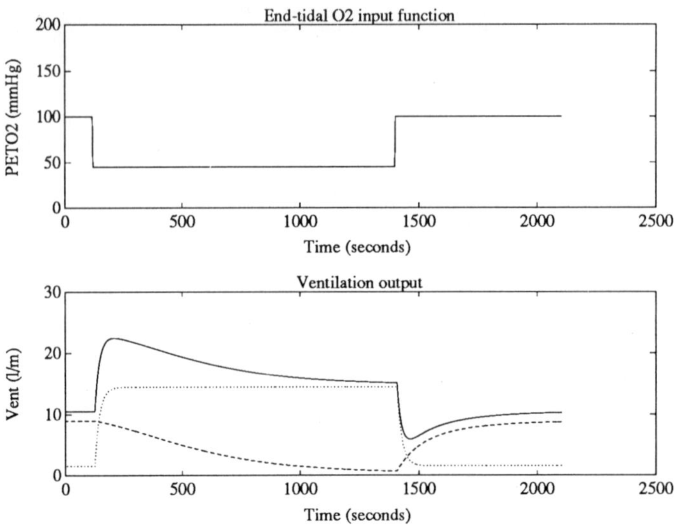

Figure 8. *Ventilatory response to a step into and out of hypoxia for the model with the peripheral component modelled with Model 1 and the central component modelled with Model 2. The model parameters are the same as in figure 2 except the value of the central time delay, Δ_d, is 9 s, identical to Δ_s. Compared with Fig. 5, note the apparent plateau in the peak ventilation without the need for a long central time delay and the asymmetric response with a much smaller undershoot during the transition out of hypoxia.*

when starting from normoxia. Sufficient experimental data is not yet available from human subjects to test this hypothesis. The asymmetry is again apparent for the reasons discussed above. The response for the step out of hypoxia is much quicker than the response to the step into hypoxia. This faster relief of hypoxic depression will result in more intermingling of the central and the peripheral responses when the composite response is modelled.

Figure 8 shows an example of the response when the peripheral response is modelled by Model 1 and the central response by Model 2. The pure time delays in the model are the same for both the central and peripheral components and are on the order of the circulation time to the brain. The use of Model 2 for the central response results in an apparent plateau in the peak response until the central depression builds up. The asymmetry in the response (compare to Fig. 5) is due to the very rapid relief of the hypoxic depression which prevents the rapid decrease in the peripheral drive from being fully seen in the ventilation.

Since the nonlinearity in the peripheral response occurs at the carotid body, the use of Model 1 for the stimulating response implies that the dynamics occurs in the neural processing. That is, the pure time delay corresponds to the circulation delay and the diffusion of oxygen out of the capillaries and into the chemoreceptor; the dynamics corresponds to the transduction of the O_2 into a neural spike train, and the nonlinearity corresponds to the neural processing of this spike train into a change in ventilation. The use of Model 2 for the central component implies a different sequence of events. Again the pure time delay corresponds to the O_2 transport from the lungs to the brain. The dynamics now correspond to a dynamic process that occurs before the nonlinearity. This could correspond to diffusion process, a depletion of oxygen stores near the central site of action or an oxygen level induced change in a neuromodulator concentration. The nonlinearity then models how the change in central tissue O_2 or the neuromodulator concentration is reflected in a change in ventilation.

At this point in model development, speculations on the model implications for physiological mechanisms should suggest experiments to confirm or deny the speculations. There may be difficulties in fitting this model to actual data, since the value of D now becomes quite critical even for step inputs. Whether or not an input can be designed that will allow the estimation of D has yet to be determined. The next step in the modelling process will be to fit this model (*i.e.*, Model 1 for the peripheral response and Model 2 for the central response) to actual data sets. Undoubtedly this process will reveal limitations in the model and the refinement process will continue.

CONCLUSIONS

The process of model building involves the repeated interplay of experiment, parameter estimation and computer simulation. Because of the nonlinear bi-phasic nature of the hypoxic ventilatory response the development of a mathematical model is not straightforward. However the building and experimental verification (or repudiation) of such models will help to understand the mechanisms of the hypoxic ventilatory response.

Acknowledgements

Many helpful discussions on modelling the hypoxic response were had with Albert Dahan, Jaap DeGoede and Aad Berkenbosch. Kamel Aqleh has assisted with the experimental work.

References

[1] van Beek, H.G.M., A. Berkenbosch, J. DeGoede, and C.N. Oliever. Effects of brainstem hypoxaemia on the regulation of breathing. *Respir. Physiol.* 57:171-188, 1984.

[2] Bellville, J.W., D.S. Ward and D. Wiberg. Respiratory System: Modelling and Identification, *in*: "Systems and Control Encyclopedia: Theory, Technology, Applications." M. G. Singh, ed., Pergamon Press, Oxford 1988.

[3] Berkenbosch, A., J. DeGoede, C.N. Oliever, J.J. Schuitmaker and D. S. Ward. Dynamics of ventilation following sudden isocapnic changes in end-tidal O_2 in cats. *J. Physiol(Lond)* 394:76P, 1987.

[4] DeGoede, J., Van Der Hoeven, N., Berkenbosch, A., Olievier, C.N. and J.H,G.M. Van Beek. Ventilatory responses to sudden isocapnic changes in end-tidal O_2 in cats. *In:* Modelling and Control of Breathing, ed. Whipp, B.J. & Wiberg, D.M., pp 37-45. Elsevier Science Publishing Co.,Inc. 1983.

[5] Dahan, A., I.C.W. Oliever, A. Berkenbosch and J. DeGoede. Modelling the dynamic ventilatory response to carbon dioxide in healthy human subjects during normoxia, *in*: "Respiratory control: Modelling perspective." G. D. Swanson and F. S. Grodins ed., Plenum, New York, in press.

[6] Easton, P.A., L.J. Slykerman and N.R. Anthonisen. Ventilatory response to sustained hypoxia in normal adults. *J. Appl. Phys.* 61:906-911, 1986.

[7] Easton, P.A., L.J. Slykerman and N.R. Anthonisen. Recovery of the ventilatory response to hypoxia in normal adults. *J. Appl. Phys.* 64:521-528, 1988.

[8] Easton, P.A. and N.R. Anthonisen. Ventilatory response to sustained hypoxia after pretreatment with aminophylline. *J. Appl. Phys.* 64:1445-1450, 1988.

[9] Easton, P.A. and N.R. Anthonisen. Carbon dioxide effects on the ventilatory response to sustained hypoxia. *J. Appl. Phys.* 64:1451-1456, 1988.

[10] Eldridge, R.L. and P. Gill-Kumar. Central neural respiratory drive and afterdischarge. *Resp. Phys.* 40:49-63, 1980.

[11] Ljung, L. "System Identification: Theory for the user." Prentice-Hall, Inc., Englewood Cliffs. 1987.

[12] Mann, C. Boetger, K. A. Aqleh and D. S. Ward. Asymmetry in the ventilatory response to a bout of hypoxia in human beings, *in*: "Respiratory control: Modelling perspective." G. D. Swanson and F. S. Grodins ed., Plenum, New York, in press.

[13] Ward, D.S., J. DeGoede, D. Wiberg, A. Berkenbosch and J.W. Bellville. Analysis of a ventilatory noise model in man and cat, *in*: "Modelling and the control of breathing." B. J. Whipp and D. M. Wiberg, ed., Elsevier Biomedical, Amsterdam, 1983.

[14] Ward, D. S., J. Degoede, A. Berkenbosch. Building dynamic models of the control of breathing during hypoxia. *in*: "Respiratory control: Modelling perspective." G. D. Swanson and F. S. Grodins ed., Plenum, New York, in press.

[15] Weil, J.V. and C.W. Zwillich. Assessment of ventilatory response to hypoxia: methods and interpretation. *Chest* 70:124-128 (Suppl), 1976.

DESIGN AND ANALYSIS OF EXPERIMENTS FOR STUDYING HYPOXIC–HYPERCAPNIC INTERACTIONS IN RESPIRATORY CONTROL

Peter A. Robbins

University Laboratory of Physiology

Parks Road, Oxford OX1 3PT, U.K.

DEFINITION OF INTERACTION

Let $x \in S_x$, where S_x is the interval of allowable hypercapnic stimuli, and let $y \in S_y$, where S_y is the interval of allowable hypoxic stimuli. A respiratory control function, θ, may be defined which maps the stimulus pair (x, y) to the steady-state ventilation, $\theta(x, y)$. For any $x_1, x_2 \in S_x$, and $y_1, y_2 \in S_y$, the stimulus changes, $\Delta x, \Delta y$, may be defined as:

$$\Delta x = x_2 - x_1$$

and

$$\Delta y = y_2 - y_1.$$

For non-interaction of the hypercapnic stimulus change with the level of hypoxic stimulation, we require:

$$\theta(x_1 + \Delta x, y_1) - \theta(x_1, y_1) = \theta(x_1 + \Delta x, y_2) - \theta(x_1, y_2).$$

For non-interaction of the hypoxic stimulus change with the level of hypercapnic stimulation, we require:

$$\theta(x_1, y_1 + \Delta y) - \theta(x_1, y_1) = \theta(x_2, y_1 + \Delta y) - \theta(x_2, y_1).$$

Both expressions, after substituting for Δx and Δy, yield:

$$\theta(x_2, y_2) + \theta(x_1, y_1) = \theta(x_1, y_2) + \theta(x_2, y_1).$$

This is the condition for non-interaction of stimuli. All other possibilities are defined as stimulus interaction.

Modeling and Parameter Estimation in Respiratory Control
Edited by M.C.K. Khoo
Plenum Press, New York

EARLY MODELS

One early model of the steady-state responses of ventilation, $\dot{V}_E$, to H^+, P_{CO_2} and P_{O_2} is that of Gray[1]:

$$\dot{V}_E(H^+, P_{CO_2}, P_{O_2}) = 0.22H^+ + 0.262P_{CO_2} - 18 + 2.118 * 10^{-8}(104 - P_{O_2})^{4.9}.$$

This type of model became known as the multiple factor theory, where pH, P_{CO_2} and P_{O_2} all contribute to $\dot{V}_E$ independently of each other— *i.e.*, the stimuli are non-interactive.

This model, however, was not in accord with the data of Nielsen and Smith[2]. These data showed clearly that the respiratory response to hypoxia became greater as the P_{CO_2} increased. To deal with this, Lloyd et al.[3] developed a model which incorporated the interaction between the hypoxic and hypercapnic stimuli:

$$\dot{V}_E = D(P_{CO_2} - B) + DA\frac{(P_{CO_2} - B)}{(P_{O_2} - C)}$$

where A, B, C and D are constants. The question arising from this model is "where does this interaction occur?".

RECEPTOR PHYSIOLOGY

There are two major sets of chemoreceptors which drive breathing, the central chemoreceptors and the peripheral chemoreceptors. The central chemoreceptors are stimulated by the acidity of the brain extracellular fluid, which in turn is determined primarily by the arterial P_{CO_2} (leaving aside any complications which may be introduced by central hypoxic depression). The peripheral chemoreceptors respond to *both* P_{CO_2} (through its effects on pH) and P_{O_2}. There are two properties of the peripheral chemoreceptors which are particularly important here:

(a) The effects of P_{CO_2} and P_{O_2} are interactive, such that the response to hypoxia is greater as the P_{CO_2} increases.

(b) Each chemoreceptor nerve fibre responds to *both* P_{CO_2} and P_{O_2}. Consequently, the information transmitted to the central nervous system is about the degree of stimulation. There is no information about whether the origins of the stimulation are primarily hypercapnic or primarily hypoxic (in the absence of any temporal coding).

Given these properties of the receptors, it is tempting to suppose that the activities of the receptors correspond with the terms of the Lloyd equation as follows:

$$\text{central chemoreceptors} \iff D(P_{CO_2} - B)$$

$$\text{peripheral chemoreceptors} \iff DA\frac{(P_{CO_2} - B)}{(P_{O_2} - C)}.$$

From these correspondences, it is clear that the next step should be to compare the nature and degree of interaction observed in ventilatory phenomena with that observed in the nervous discharge from the peripheral chemoreceptors. In man, the increase in ventilation observed with a given hypoxic stimulus is around 3–4 times greater at a P_{CO_2} of 50 Torr than it is at a P_{CO_2} of 40 Torr[4]. In the anesthetised cat, the increment in firing rate with a given hypoxic stimulus is only one-third greater at a P_{CO_2} of 50 Torr than it is at a P_{CO_2} of 40 Torr[5]. The most obvious explanation for this quantitative difference must be species difference. However, problems arise even if we take only the qualitative features of the chemoreceptor response as applying to man. First, the chemoreceptor response is reasonably linear with P_{CO_2}; secondly, the slope of this relation increases with increasing hypoxia; and thirdly, the P_{CO_2} at which the chemoreceptors begin to discharge falls with increasing hypoxia. A consequence of these features is that the lines describing the response of the chemoreceptors to P_{CO_2} for different levels of hypoxia can only intersect if they are extrapolated to negative nerve impulse frequencies. Applying these qualitative ideas to the ventilatory data of man means that the chemoreceptors should be silent at and below the P_{CO_2} at which any of the ventilation–P_{CO_2} lines meet when extrapolated backwards. Inspection of the steady-state data, however, reveals that hypoxia, presumably acting via the peripheral chemoreceptors, drives ventilation to levels of P_{CO_2} below the intersection points, where the peripheral chemoreceptors should be silent. Unfortunately, the striking nature of this inconsistency is to some extent lost when the argument is complicated with the central depressant effects of hypoxia. Nevertheless, the argument still suggests that the interactions between hypercapnia and hypoxia in the ventilatory responses of man cannot be explained solely in terms of the interactions observed in the discharge rates of the peripheral chemoreceptors.

EXPERIMENTS ON ANESTHETISED ANIMALS

There have been a number of experiments on anesthetised animals, and a range of conclusions have been drawn. Perhaps the most compelling evidence comes from experiments using an artificial brainstem perfusion technique. Van Beek et al.[6] perfused the brainstem separately from the rest of the animal. In this way they were able to examine the central CO_2 sensitivity at different levels of peripheral stimulation. They found the central CO_2 sensitivity was unchanged by the level of peripheral stimulation, and concluded that all the interaction between CO_2 and hypoxia occurred peripherally. They considered a non-interactive model was sufficient to describe the ventilation resulting from a combination of peripheral and central drives.

EXPERIMENTS ON HUMAN VOLUNTEERS

Taken on their own, the steady-state experiments on humans which have demonstrated interaction between hypoxic and hypercapnic stimuli give no indication where that interaction may occur. Furthermore, experiments on anesthetised animals that require physical separation of the blood supply to the peripheral and central chemoreceptors are clearly impossible in humans. Experiments on humans have instead relied on dynamic techniques to localize the site of interaction by exploiting the difference in the speeds of response of the peripheral and central chemoreflex loops. Miller et al.[7] examined the effect of withdrawing a hypercapnic stimulus against a background of either hypoxia or of hyperoxia. They found that that the first decrease in $\dot{V}_E$ was after 2–3 breaths in hypoxia, but only after 5 breaths in hyperoxia. The interpretation they placed on this result was that the shorter latency response was due to the peripheral chemoreflex loop, and that the response to CO_2 at the peripheral chemoreceptor was abolished by hyperoxia— *i.e.*, interaction occurs between CO_2 and hypoxia at the peripheral chemoreceptors. The experiment does *not*, however, provide evidence as to whether any additional interaction occurs between the peripheral and central chemoreflex loops.

The question of whether there is any additional interaction between the peripheral and central chemoreflex loops in man is somewhat harder to answer. Robbins[8] used a dynamic technique which employed step changes in both alveolar P_{CO_2} and P_{O_2} to assess whether any peripheral-central interaction occurred. The end-tidal P_{CO_2} was initially held at 50 torr and the end-tidal P_{O_2} held at 100 torr until ventilation was relatively stable. The end–tidal P_{CO_2} was then reduced in a steplike manner to 40 torr. This produces a temporal separation of the P_{CO_2} at the central and peripheral chemoreceptors, such that after 30 s the P_{CO_2} at the peripheral chemoreceptors will have adjusted to around the new level, whereas the P_{CO_2} at the central chemoreceptors will still be considerably raised (because the dynamics of equilibration of CO_2 are so much slower centrally). Against this background, the end-tidal P_{O_2} was reduced in a steplike manner to 50 Torr. The idea was that, if all the interaction occurs peripherally, then the raised central P_{CO_2} should make no difference to the hypoxic response. Thus the hypoxic response observed should be the same as that observed when the lower P_{CO_2} is fully equilibrated at *both* the peripheral and central chemoreceptors. On the other hand, if there is interaction between the peripheral and central chemoreflex loops, then the hypoxic response should be greater when the central P_{CO_2} is raised than when it is normal. Robbins[8] found that the hypoxic response was indeed augmented in two out of three of his subjects with the raised central P_{CO_2}.

PREDICTIONS ARISING FROM AN INTERACTIVE MODEL

In this section a simple interactive model for the ventilatory controller will be considered. The dynamics of the peripheral and central chemoreceptors will be modeled using two simple linear first-order differential equations. The total output will be taken as the sum of the outputs of the two loops, plus an additional term depending on the product of the two outputs. The model may be written:

$$\tau_c \frac{di_c}{dt} + i_c = g_c(P_{CO_2}(t) - c_c)$$

$$\tau_p \frac{di_p}{dt} + i_p = g_p(P_{CO_2}(t) - c_p)$$

$$i_t = i_c + i_p + g_i(i_c i_p)$$

where i_c and i_p are the central and peripheral outputs, τ_c and τ_p are the central and peripheral time constants, g_c and g_p are the central and peripheral gain terms, c_c and c_p are the central and peripheral bias terms, i_t is the total respiratory output, and g_i is the interactive gain term.

Sinusoidal Inputs

With a sinusoidal CO_2 input of angular velocity ω, the output responses of the individual loops have the form:

$$i_c = a_c \sin(\omega t + \phi_c) + b_c$$

$$i_p = a_p \sin(\omega t + \phi_p) + b_p$$

where a_c and a_p are the amplitudes of oscillation of the central and peripheral chemoreceptor outputs, ϕ_c and ϕ_p are the phase shifts, and b_c and b_p are the mean levels of the outputs for the central and peripheral chemoreflex loops. If there were no interaction between the outputs of the loops, then the total output would just be a sum of sine waves. However, the interaction in the model adds the term:

$$g_i((a_c \sin(\omega t + \phi_c) + b_c)(a_p \sin(\omega t + \phi_p) + b_p).$$

This gives on multiplying out an expression of the form:

$$\alpha + \beta \sin(\omega t + \phi_1) + \gamma \sin(2\omega t + \phi_2)$$

where $\alpha, \beta, \gamma, \phi_1$ and ϕ_2 are constants. The most interesting result here is the appearence of a term in the output of twice the frequency of the input. This result suggests that an analysis of the response to a sinusoid of P_{CO_2} at constant P_{O_2} may provide evidence for interaction. However, this has not been followed up for two reasons. First, further modeling suggests that the likely magnitude of the first harmonic would be too small to be detectable, and secondly, the effect is too non-specific in that non-linearities in the response of the chemoreceptors themselves could also give rise to this effect.

Step Inputs

With a step increase in P_{CO_2} at constant P_{O_2}, the output responses of the individual loops have the form:

$$i_c = a_c + k_c(1 - \exp(-t/\tau_c))$$

$$i_p = a_p + k_p(1 - \exp(-t/\tau_p))$$

where a_c, k_c, a_p and k_p are constants. If there were no interaction between the outputs of the loops, then the total output would just be the sum of two exponentials. However, with the interactive model, there is also a term proportional to the product of the two right-hand sides, which give on multiplying out:

$$a_c a_p + a_c k_p(1 - \exp(-t/\tau_p)) + a_p k_c(1 - \exp(-t/\tau_c)) + k_c k_p(1 - \exp(-t/\tau_c))(1 - \exp(-t/\tau_p)).$$

Multiplying out the last term gives, for that term:

$$k_c k_p(1 - \exp(-t/\tau_p) - \exp(-t/\tau_c) + \exp(-t/\tau_p)\exp(-t/\tau_c)).$$

Now, assuming $\tau_p \ll \tau_c$, the last term in this expression involving the product of the exponentials may be approximated by $\exp(-t/\tau_p)$ and the expression reduces to

$$k_c k_p(1 - \exp(-t/\tau_c)).$$

The entire interactive component may be written:

$$a_c a_p + a_c k_p(1 - \exp(-t/\tau_p)) + a_p k_c(1 - \exp(t/\tau_c)) + k_c k_p(1 - \exp(-t/\tau_c)).$$

For a step decrease in P_{CO_2}, the output responses for the central and peripheral chemoreceptors have the form:

$$i_c = a_c + k_c\exp(-t/\tau_c)$$

$$i_p = a_p + k_p\exp(-t/\tau_p).$$

The product is:

$$a_c a_p + a_c k_p\exp(-t/\tau_p) + a_p k_c\exp(-t/\tau_c) + k_c k_p\exp(-t/\tau_c)\exp(-t/\tau_p).$$

Assuming $\tau_p \ll \tau_c$, the term involving the product of exponentials approximates to $k_c k_p\exp(-t/\tau_p)$. The entire interactive component may be written:

$$a_c a_p + a_c k_p\exp(-t/\tau_p) + a_p k_c\exp(-t/\tau_c) + k_c k_p\exp(-t/\tau_p).$$

An examination of the interactive components for the on- and off-transients yields a number of interesting features. First the on- and off-transients are asymmetric. The component of gain $k_c k_p$ has central chemoreceptor dynamics (time constant τ_c) for the on-transient, and peripheral chemoreceptor dynamics (time constant τ_p) for the off-transient. Unfortunately, this feature has not been particularly helpful in determining whether there is any interaction. First, a degree of asymmetry in the values for the central time constants for the on- and off-transients has been reported, which confuses the analysis, and secondly, the gain terms have in general been fitted as single values for the on- and off-transients combined[9,10].

A second interesting feature of the interactive component is the appearance of a term with central chemoreceptor dynamics (time constant τ_c) and gain term $a_p k_c$ in both the on- and off-transients. The reason that this term is interesting is that, for a given step change in end-tidal P_{CO_2}, its magnitude depends on the degree of initial peripheral chemoreceptor activity, a_p; the greater this initial activity the greater will be the overall gain of the term with central chemoreceptor dynamics. This leads to the prediction that, for an interactive model, the gain of the component with central chemoreceptor dynamics (τ_c) would increase with increasing levels of hypoxia, whereas for a non-interactive model, the gain should be independent of the level of hypoxia. This prediction can be tested against data already in the literature. Examining the data of Bellville et al.[10] reveals that six out of seven of their subjects showed a higher central chemoreceptor gain in hypoxia than in hyperoxia, and furthermore the central chemoreceptor gain of subjects who had undergone carotid body resection ($a_p = 0$) was around half that of normal subjects. This very encouraging finding is tempered by the data of Ward and Bellville[11], which show no reduction of central chemoreceptor gain with infusions of dopamine despite a significant reduction in peripheral chemoreceptor gain.

CONCLUSIONS

Interaction exists between the respiratory stimuli of hypercapnia and hypoxia. Some of this interaction clearly arises at the level of the peripheral chemoreceptor in both the anesthetised cat and in conscious man. In the anesthetised cat, experiments involving individual control of the chemical environment of the peripheral and central chemoreceptors suggest that no interaction is required between the central and peripheral chemoreceptors to explain the overall ventilatory responses. There are, however, clearly difficulties in attributing all the hypoxic-hypercapnic interaction observed in man to peripheral chemoreceptors with the qualitative properties of those of the anaesthetised cat. There are a number of predictions about the nature of the ventilatory responses to dynamic stimuli which arise if central-peripheral interaction occurs in man. Some of these have been explored, and in some cases evidence for central-peripheral interaction has been forthcoming. However, it is as yet too soon for any firm conclusions to be drawn about the presence or absence of this phenomenon in conscious humans.

REFERENCES

1. J.S. Gray, "Pulmonary Ventilation and its Physiological Regulation," Thomas, Springfield (1950).

2. M. Nielsen and H. Smith, Studies on the regulation of respiration in acute hypoxia, Acta Physiol. Scand. <u>24</u>:293–313 (1952).

3. B.B. Lloyd, M.G.M. Jukes, and D.J.C. Cunningham, The relation between alveolar oxygen pressure and the respiratory response to carbon dioxide in man, Quart. J. Exp. Physiol. <u>43</u>:214–227 (1958).

4. B.B. Lloyd and D.J.C. Cunningham, Quantitative approach to the regulation of human respiration, <u>in</u>: "The Regulation of Human Respiration," D.J.C. Cunningham and B.B. Lloyd, ed., Blackwell, Oxford (1963).

5. S. Lahiri and R.G. Delaney, Stimulus interaction in the responses of carotid chemo-receptor single afferent fibres, Respir. Physiol. 24:249–266 (1975).

6. J.H.G.M. van Beek, A. Berkenbosch, J. de Goede, and C.N. Olievier, Influence of peripheral O_2 tension on the ventilatory response to CO_2 in cats, Respir. Physiol. 51:379–390 (1983).

7. J.P. Miller, D.J.C. Cunningham, B.B. Lloyd, and J.M. Young, The transient respiratory effects in man of sudden changes in alveolar CO_2 in hypoxia and in high oxygen, Respir. Physiol. 20:17–31 (1974).

8. P.A. Robbins, Evidence for interaction between the contributions to ventilation from the central and peripheral chemoreceptors in man, J. Physiol. (London) 401:503–518 (1988).

9. G.D. Swanson and J.W. Bellville, Step changes in end-tidal CO_2: methods and implications, J. Appl. Physiol. 39:377–385 (1975).

10. J.W. Bellville, B.J. Whipp, R.D. Kaufman, G.D. Swanson, K.A. Aqleh, K.A. and D.M. Wiberg, Central and peripheral chemoreflex loop gain in normal and carotid body-resected subjects, J. Appl. Physiol. 46:843–853 (1979).

11. D.S. Ward and J.W. Bellville, Effect of intravenous dopamine on hypercapnic ventilatory response in humans, J. Appl. Physiol. 55:1418–1425 (1983).

ESTIMATION OF DYNAMIC CHEMOREFLEX GAIN FROM SPONTANEOUS
BREATHING DATA

Michael C.K. Khoo

Biomedical Engineering Dept., University of Southern
California, Los Angeles, CA 90089

INTRODUCTION

The chemoreflex control of ventilation is traditionally assessed by presenting the respiratory system with different inhalation mixtures of hypercapnic and/or hypoxic gas. In the steady state method[1], the composition of the inhalate is kept constant and the "steady state" ventilatory response is measured several minutes after the start of the test. In the rebreathing method introduced by Read[2], the stimulus increases progressively with time in a ramp-like function. However, from these simple techniques, we can only deduce the <u>combined steady state</u> responses of the central and peripheral chemoreceptors. The <u>dynamic</u> characteristics of the chemoreflexes cannot be inferred from such tests. It is generally accepted that the breath–to–breath control of ventilation is achieved primarily through the reflex action of the fast–responding peripheral chemoreceptors[3]. Interventions which augment peripheral chemosensitivity have been shown in theoretical[4] and experimental[5,6] studies to produce periodic breathing or ventilatory oscillations. In anesthetized animals, it is possible to achieve physical separation of the central and peripheral drives by isolating the perfusion of the brain–stem from that of the carotid bodies[7]. Such invasive procedures are clearly impossible to apply in humans. Consequently, the tests of dynamic respiratory control that are performed in humans have been limited to techniques in which rapid changes in inspired gas concentrations are effected. These techniques are based on the premise that the two groups of chemoreceptors can be <u>functionally</u> separated, since the central chemoreflex is believed to be considerably more sluggish than the peripheral chemoreflex. The responses to hypercapnic steps, pulses and sine–waves have been examined by several investigators[8–12].

A common problem encountered in all these methods is that the natural breath–to–breath variation in ventilation, particularly in conscious humans, can be quite substantial. In order to improve the 'signal–to–noise ratio', reasonably potent levels of test stimuli have to be employed. In many cases, the subject can easily taste or become aware of the non–physiologic stimulus. In some, the cough reflex may be triggered. Furthermore, the system is never actually observed in its normal operating state but is perturbed during the measurement process by these abnormal levels of test stimuli. Thus, strictly speaking, the CO_2 inhalation tests provide information only about CO_2 sensitivity in the region <u>above</u> eucapnia. Experiments on anesthetized cats have demonstrated a marked steepening of the CO_2 response slope below the eucapnic point[13]. In man, the question as to whether there is a flattening (i.e. a "dog–leg effect") or a steepening of the CO_2 response remains controversial[3]. In the latter situation, a measure of the hypercapnic sensitivity would lead us to overestimate the degree of stability of the respiratory control system.

The natural variations in breathing pattern include occasional sighs. Sighs represent impulse–like disturbances in ventilation that produce transient changes in alveolar P_{CO2} (P_{ACO2}) which, in turn, lead to subsequent changes in minute ventilation ($\dot{V}_E$). Post–sigh ventilatory responses have been used to characterize the degree of stability in respiratory control[14-16]. We have shown in a recent theoretical and experimental study that a reasonably good index of peripheral chemosensitivity can be obtained by analyzing the changes in $\dot{V}_E$ and end–tidal P_{CO2} that follow a sigh[17]. Aside from occasional sighs, the spontaneous variations which occur in the breathing patterns of adult humans are generally composed of oscillations with distinct periods as well as random fluctuations of broad frequency content[18,19]. The problem addressed in this chapter is whether careful analysis of these spontaneous variations in $\dot{V}_E$ and P_{ACO2} can yield useful quantitative information about the dynamic characteristics of the chemoreflexes. Since these perturbations are purely physiologic and include deviations above and below the eucapnic point, the measure of dynamic chemosensitivity that results is truly representative of the "impulse response" of the respiratory controller under normal operating conditions. In the discourse that follows, we begin by presenting the theoretical background behind our new method of analysis. The method is next tested with 'data' simulated by a model of the closed–loop respiratory control system. Finally, we apply the estimation procedure to experimental data.

METHOD OF ANALYSIS – THEORY

There are a number of important assumptions in the present method of analysis:

(1) As in most other studies of ventilatory response, we assume that end–tidal P_{CO2} provides an accurate representation of P_{ACO2}. During quiet periods of resting ventilation, this is a reasonable assumption to make[20]. However, corrections would have to be made for the occasional shallow breaths that are not much larger than the dead space, in which case end–tidal P_{CO2} would not accurately represent P_{ACO2}.

(2) We assume that the CO_2 controller is linear, at least within the range of fluctuations in $\dot{V}_E$ and end–tidal P_{CO2}.

(3) The closed–loop respiratory control system is perturbed continually by a random broad–band extraneous ventilatory drive. The dynamic response of the chemoreflexes to these perturbations gives rise to resonances which take the form of narrow–band oscillations. Thus, the total ventilatory drive at any breath is the sum of the non–random chemoreflex drive and the random extraneous drive.

(4) The $\dot{V}_E$ and end–tidal P_{CO2} time–series used in the analysis are stationary.

(5) The time taken for changes in the lungs to first appear at the site of the peripheral chemoreceptors is N_D (in units of breaths). Under resting conditions, we would expect N_D to range from 1 to 3 breaths.

(6) The calculations assume a time–scale based on number of breaths. Thus, implicit in this assumption is the supposition that the breath duration remains constant from breath to breath.

We will represent by x and y, the respective fluctuations in P_{ACO2} (or end–tidal P_{CO2}) and $\dot{V}_E$ about their corresponding means. Representing the random extraneous drive at breath k as $e(k)$, and the impulse response of the controller as $h(k)$, the change in $\dot{V}_E$ at the current breath is related to past values of changes in P_{ACO2} by:

$$y(k) = e(k) + \sum_{i=o}^{\infty} h(i).x(k - i - N_D) \tag{1}$$

We define the cross–correlation between x and y, where y lags x by m breaths, in the following manner:

$$R_{xy}(m) = \frac{1}{N} \sum_{i=o}^{N-m-1} x(i).y(i + m), \quad 0 \leq m \leq M \tag{2}$$

where N is the total number of breaths in the sequence and M is the maximum lag used in the correlation computation, $M << N$. Cross–correlating both sides of Eq. 1 with x, we obtain:

$$R_{xy}(m) = R_{xe}(m) + \sum_{i=o}^{\infty} h(i).R_{xx}(m - i - N_D) \tag{3}$$

Suppose the impulse response of the closed–loop respiratory control system is $f(k)$. Then, since $x(k)$ will be the response of the system to the extraneous disturbance $e(k)$, i.e.:

$$x(k) = \sum_{i=-\infty}^{\infty} f(i).e(k - i) \tag{4}$$

From Eq. 4, we can deduce that:

$$R_{xe}(m) = \sum_{i=-\infty}^{\infty} f(i).R_{ee}(m + i) \tag{5a}$$

$$= \sum_{i=-\infty}^{\infty} f(-i).R_{ee}(m - i) \tag{5b}$$

If $e(k)$ is white noise (i.e. uncorrelated with itself, except at $m = 0$), $R_{ee}(m)$ will take the form of a delta function, and consequently,

$$R_{xe}(m) \propto f(-m) \tag{6}$$

Since the system is causal, $f(m) = 0$ for $m < 0$. Thus, in Eq. 6:

$$R_{xe}(m) = 0, \quad m > 0 \tag{7}$$

Thus, if $e(k)$ is white, Eq. 3 becomes:

$$R_{xy}(m) = \sum_{i=o}^{\infty} h(i).R_{xx}(m - i - N_D), \quad 0 < m \leq M \tag{8}$$

The above result may be extended to the more general case where $e(k)$ is broad–band (i.e. correlated with itself only at the first few (e.g. up to m_0, where $m_0 > 0$) lags):

$$R_{xy}(m) = \sum_{i=o}^{\infty} h(i).R_{xx}(m - i - N_D), \quad m_0 < m \leq M \tag{9}$$

The approach we have adopted for estimating the controller impulse response in Eq. 9 is as follows. First, we assume a simple but realistic form for $h(i)$:

$$h(i) = \sum_{q=1}^{N_c} (G_q/\tau_q).exp(i/\tau_q), \quad i \geq 0 \tag{10}$$

where N_c is the maximum number of exponential functions that $h(i)$ is composed of. Substituting Eq. 10 into Eq. 9, and truncating the infinite sum beyond the $(p+1)$th term:

$$R_{xy}(m) = \sum_{i=o}^{P} \sum_{q=1}^{N_c} (G_q/\tau_q).exp(i/\tau_q).R_{xx}(m - i - N_D), m_0 < m \leq M \tag{11}$$

Note that in Eq. 11, the following parameters are unknown: m_0, N_D, G_q and $\tau_q(q = 1,\ldots,N_c)$. The estimation of m_0 is described later. For given values of these unknown parameters, we compute from Eq. 11 "predicted" values for $R_{xy}(m)$ at all feasible m, using values of R_{xx} deduced from the data. These are compared to the corresponding "observed" values for $R_{xy}(m)$, computed from the data using Eq. 2. We define a criterion function, J, to quantify the discrepancy between the predicted and observed values:

$$J = \sum_{m=m_0+1}^{M} \left(R_{xy}(m)_{\text{predicted}} - R_{xy}(m)_{\text{observed}} \right)^2 \tag{12}$$

Using an optimization algorithm[20], we search for the set of values of G_q and $\tau_q(q = 1,\ldots,N_c)$ that minimizes J, given a particular value of N_D. The optimization process is repeated for other feasible values of N_D. The combination of parameters that produces the global minimum in J is accepted as the best estimate of N_D, G_q and $\tau_q(q = 1,\ldots,N_c)$. An alternative method for computing $h(i)$ has been described elsewhere[21].

TESTS WITH SIMULATED DATA

The Model

It is useful to first apply the estimation procedure to 'data' simulated by a respiratory control model. Since the exact parameter values employed in the model are known, it is possible to check the estimates obtained with our procedure against the corresponding 'true' values. This ensures that any discrepancy that appears is not the result of experimental or measurement error.

The model used for generating the 'data' has been described in detail in Khoo and Marmarelis[17]. Briefly, CO_2 exchange between alveolar gas and pulmonary capillary blood was assumed to take place in a homogeneous fixed–volume, flow–through chamber. A fraction of the total ventilation did not participate in the gas exchange process but instead flowed through an unperfused chamber (the dead space). Time delays were included to account for the transport of blood from the chamber to the chemoreceptors. First–order differential equations, similar to those in the model of Swanson and Bellville[22], were used to describe the dynamic responses of the central and peripheral chemoreflexes. Total ventilatory drive were assumed to be the sum of the central and peripheral drives, plus an extraneous Gaussian white noise component. The actual ventilation in any given breath was determined by the value of total ventilatory drive appearing at the <u>start</u> of that breath; during the breath, this value of ventilation was kept constant.

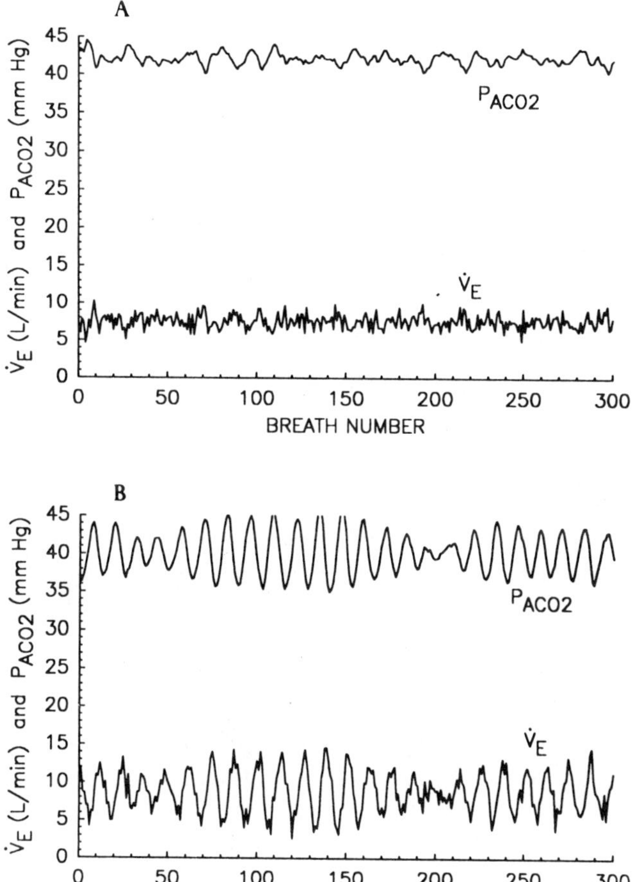

Fig. 1.   Breath–to–breath values of V_E and P_{ACO2} generated by the closed–loop model of respiratory control. (A) Case with low peripheral gain (0.4 l/min/torr). (B) Case with high peripheral gain (1.5 l/min/torr).

Simulations were performed using values of central chemoreflex gain (G_c) that ranged from 0 to 2 liters/min/torr, and values for peripheral chemoreflex gain (G_p) ranging from 0 to 1.5 liters/min/torr. The time constants associated with the central chemoreflex (τ_c) and peripheral chemoreflex (τ_p) were 102 and 10 seconds, respectively. Breath duration was assumed constant at 3 seconds. The lung–to–carotid body and lung–to–medulla delays were assumed to be 6 and 7 seconds, respectively. In all cases, the standard deviation of the extraneous white noise input was 1 liter/min. Two examples of the breath–to–breath $\dot{V}_E$ and P_{ACO2} time series generated by the closed–loop model, with G_c set equal to 1 liter/min/torr, are shown in in Figs. 1A and 1B. In Fig. 1A, the peripheral gain is low ($G_p = 0.4$), and thus $\dot{V}_E$ is dominated by the random extraneous input. On the other hand, in the case of high peripheral gain ($G_p = 1.5$ in Fig. 1B), the response of the closed–loop system to the extraneous noise is highly oscillatory; notice that the interaction between the random and non–random components gives rise to <u>bursts</u> of oscillations. The oscillations have an average cycle time of 37 seconds, showing that they are mediated by the peripheral chemoreflex loop[4].

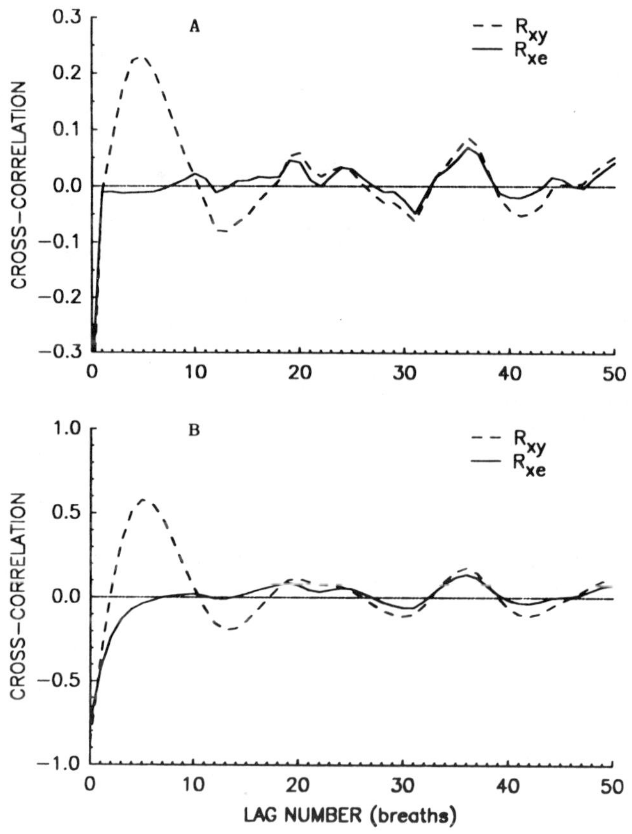

Fig. 2.   Variation of the cross–correlations, R_{xe} (solid curve) and R_{xy} (dashed curve), with number of lags (breaths), when the extraneous noise, e, is: (A) white; (B) colored.

<u>Validity of Equation (9)</u>

The validity of Eqn. 9, which forms the basis of our estimation procedure, depends on the assumption that R_{xe}, the cross–correlation between current values of x and later values of e, approximates zero after a small number of lags (m_0). To determine whether this assumption may be made, we examine in Fig. 2 how R_{xe}, computed from simulated data, varies with lag number. In Case A, e is white noise. As such, R_{xe} fluctuates around zero at all positive lags. The reason is that the extraneous noise input affects ventilation which, in turn, affects P_{ACO2} in the <u>same</u> breath. Thus, e is negatively correlated with x in the same breath (i.e. at zero lag). However, since x does not have any effect on future values of e, R_{xe} is essentially zero at all positive lags. In Case B, e is colored noise, and therefore, current values of e are correlated with past and future values. This means that there is some correlation between current values of e and past values of x (since the latter were dependent on past values of e). Consequently, R_{xe} tends towards zero only after a finite number ($m_0 = 6$) of positive lags. It should be noted that, for lags greater than m_0, R_{xe} fluctuates around zero but is not exactly zero. This behavior is a consequence of the finite length (300 breaths) of the data segment used in the calculations. A further point to note from Fig. 2 is that R_{xy} behaves essentially like R_{xe} at lags 20 and above, which implies that the first 20 lags contains basically all the information about controller dynamics that one can extract from the data.

<u>Results of the Estimation Procedure</u>

An illustration of a typical result derived from our computational procedure is shown in Fig. 3. From the model–simulated time series of P_{ACO2} and $\dot{V}_E$ shown in Fig. 3A, we calculated R_{yy} (Fig. 3B), R_{xx} (dashed line in Fig. 3C) and R_{xy} (solid line in Fig. 3C). Using Eqn. 9, we obtained the best least–squares fit (shown as the dot–dashed line in Fig. 3C) to R_{xy}. In Fig. 3B, it can be seen that R_{yy} resembles R_{ee} very closely at lags 0 and 1, as well as above lag 20. We have also found that, in cases where e is colored, R_{yy} tracks R_{ee} closely up to lag m_0 (not shown). These observations suggest that a practical means of estimating m_0 is simply to determine the lag corresponding to the first zero–crossing of R_{yy}.

We found that a single exponential function (i.e. $N_c = 1$) was sufficient for describing the impulse response of the controller. Increasing N_c to 2 and above did not result in significant improvement of the least–squares residuals. Moreover, in most cases, parameter variances became very large and the optimization algorithm frequently failed to converge. Attention was therefore restricted to the single–exponential case.

For a given combination of G_c and G_p, 10 different pairs of $\dot{V}_E$ and P_{ACO2} time–series were obtained using a different white–noise realization in each case. From each pair of $\dot{V}_E$ and P_{ACO2} time–series, we deduced estimates of Gp (G_{est}) and τ_p (τ_{est}). The mean and standard deviation were calculated from the ten estimates of each parameter.

In Fig. 4, G_{est} (mean $\pm$ standard deviation) is plotted against G_p. The line of identity is shown as a dashed line. The closed circles represent the cases in which G_c was set equal to zero, whereas for the closed triangles, G_c was set equal to 1 liter/min/torr. The effect of increased central gain is to produce a bias in G_{est}, but the slope of the linear dependence of G_{est} on G_p remains unchanged. Increasing G_c to 2 liters/min/torr (with G_p kept constant at 0.5 liter/min/torr) increases G_{est} further (top open square). The influence of central gain (in the case where $G_c = 1$) is decreased when the peripheral time constant (τ_p) is halved (lower open square, Fig. 4).

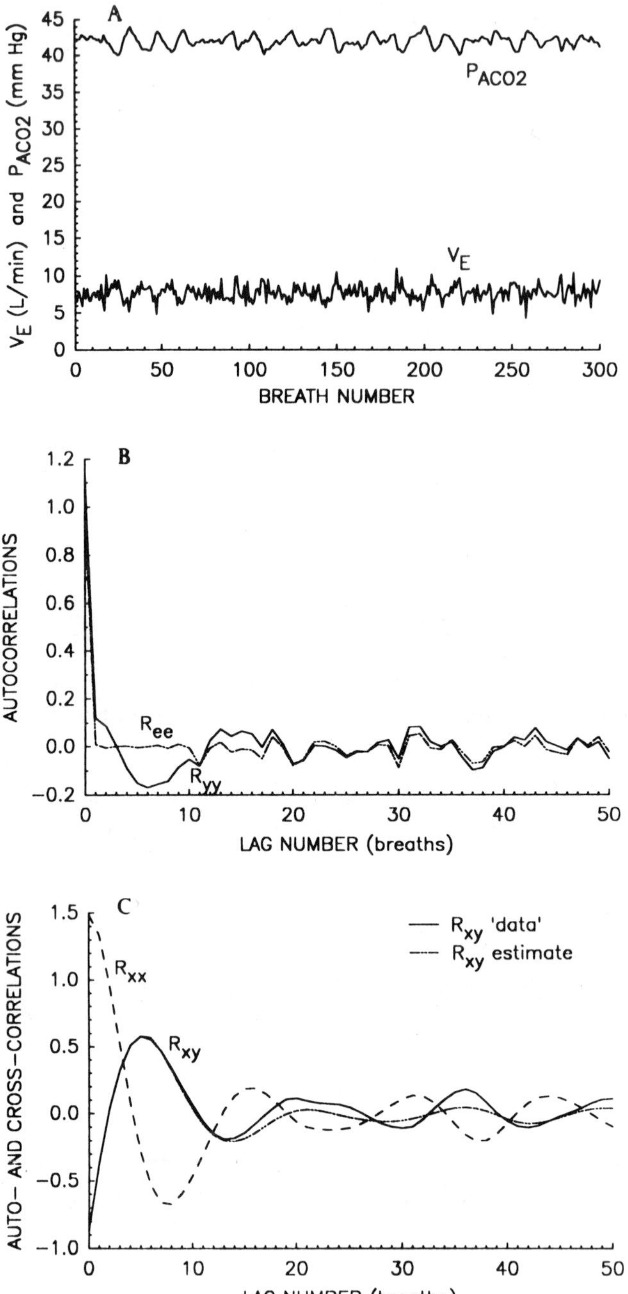

Fig. 3. An example of the results derived from applying the estimation procedure to simulated data. (A) The model–generated V_E and P_{ACO2} time series. (B) Autocorrelations, R_{ee} and R_{yy}. (C) Input autocorrelation R_{xx} (dashed curve), and cross–correlation, R_{xy}. The solid curve represents R_{xy} computed directly from the 'data', while the broken–line tracing represents best–fit prediction of R_{xy}.

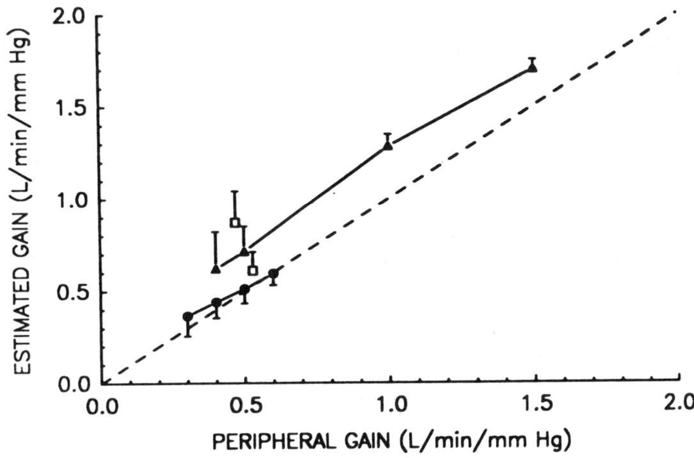

Fig. 4. Gain (means ± SD) estimated from simulated data plotted against actual peripheral gain. Circles: $G_c = 0$. Triangles: $G_c = 1$. Top square: $G_c = 2$. Lower square: $G_c = 1$ but τ_p is halved. Dashed line represents line of identity.

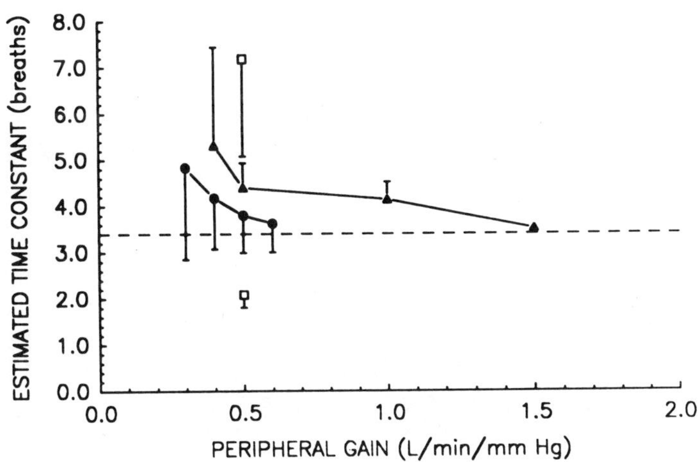

Fig. 5. Estimated time constant versus peripheral gain. Dashed line represents true value of τ_p. Symbols as defined in Fig. 4.

The influence of the central chemoreflex on estimates, τ_{est}, of the peripheral time constant is demonstrated in Fig. 5. The dashed line represents the value of τ_p employed in the simulations. As in Fig. 4, the cases for $G_c = 0$ and $G_c = 1$ are displayed as closed circles and closed triangles, respectively. As G_p decreases, the standard deviations of τ_{est} increase, and the mean value of τ_{est} overestimates τ_p by progressively larger amounts. The presence of the central chemoreflex also leads to overestimates of τ_p. With a central gain of 2 l/min/torr, mean τ_{est} is more than twice as large as the true τ_p (top open square). Thus, the biasing effect of the central chemoreflex on estimates of τ_p is substantially greater than that on estimates of G_p.

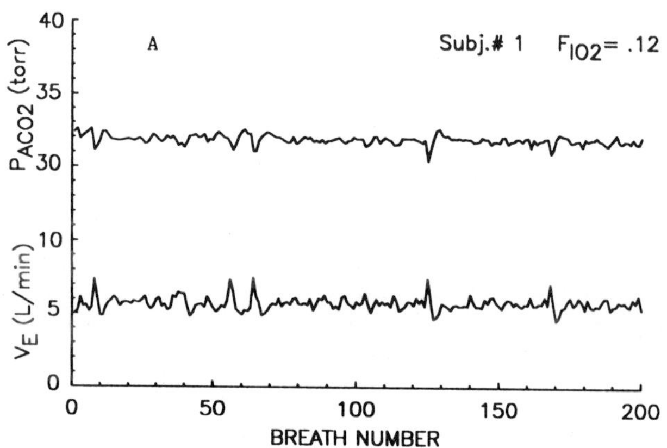

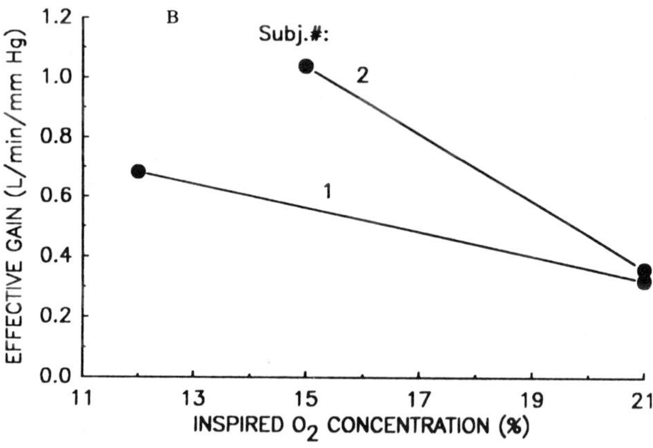

Fig. 6.   (A) Example of spontaneous breathing record from an anesthetized dog inspiring hypoxic gas (12% O_2 in N_2). (B) Gains estimated from the 2 dogs under normoxia and hypoxia are plotted against level of oxygenation. Dynamic gain is increased twofold or more by hypoxia.

TESTS WITH EXPERIMENTAL DATA

Experimental Protocol

Since differences in the degree of awareness can complicate measurements of spontaneous breathing in conscious humans, we applied the estimation procedure to respiratory data obtained from anesthetized dogs.

Spontaneous breathing was monitored in two mongrel dogs, anesthetized with Brevital Sodium (loading dose of 8 mg/kg, with subsequent infusion rate of 13 mg/kg/hr) and intubated with a cuffed endotracheal tube. Airflow, measured by means of a pneumotachometer (Rudolph 3700 with Validyne DP–45 pressure transducer), was electronically integrated (Validyne FV156–871) to obtain tidal volume. End–tidal P_{CO2} was sampled and analyzed using a Beckman LB–2 CO_2 analyzer. Blood samples could be withdrawn for subsequent blood–gas analysis and arterial blood pressure was monitored through a catheter inserted into the femoral artery. The level of anesthesia was adjusted so that the animals were in a relatively constant plane in the moderate–to–deep state[23]. Measurements of spontaneous respiration, with air as the inspirate, were recorded when the animals showed relatively stable levels of arterial blood pressure, heart rate and breathing rate. Subsequently, the inspired gas was changed to a hypoxic mixture (12% O_2–88% N_2 for Dog 1, 15% O_2–85% N_2 for Dog 2). Recording of spontaneous breathing did not commence until arterial PO_2 stabilized to a steady level, and arterial blood pressure, heart rate and respiratory rate became reasonably constant.

Results of Estimation Procedure

An example of the measurements obtained during hypoxia is shown in Fig. 6A. Although it is difficult to distinguish from the raw data, the autocorrelations (R_{xx} and R_{yy}) in Fig. 7A clearly demonstrate the existence of a sustained oscillation with period of 6–7 breaths (or approximately 21 seconds). This is also clearly shown in Fig. 7B by the form of R_{xy} (solid line). Using our estimation procedure, the gain and corresponding time–constant were deduced from the best–fit to R_{xy} (dot–dashed line, Fig. 7B). The complete results for estimated peripheral gain are displayed in Fig. 6B. In both subjects, hypoxia increased estimated gain two– to three–fold, which is what one would expect. This result shows that our gain estimates are sensitive to changes in peripheral chemoreflex gain that occur within the physiological range. However, since the actual change in CO_2 sensitivity was not known, we cannot infer from this result alone how accurate our estimation procedure is.

LIMITATIONS

It is clear that the estimation procedure we have proposed can provide a useful noninvasive means for quantifying the dynamic characteristics of the chemoreflexes, primarily the peripheral chemoreflex. However, we face a major obstacle in attempting to validate this procedure experimentally: i.e. the difficulty of finding a reliable and independent basis for comparison. One possibility that immediately comes to mind would be to check our estimates against CO_2 sensitivities deduced from steady state or rebreathing measurements taken during normoxia and hypoxia. But this would lead us back to the original problem, because such measurements only provide information about CO_2 sensitivity in the <u>hypercapnic</u> range but not in the vicinity of eucapnia.

It could be argued that alternative tests, such as the single–breath or pseudorandom binary CO_2 inhalation techniques, produce transient increases in arterial P_{CO2} that are not as large and are therefore more physiologic. However, even in these tests, the stimulus is still an increase in P_{aCO2} above eucapnia: it is highly conceivable that

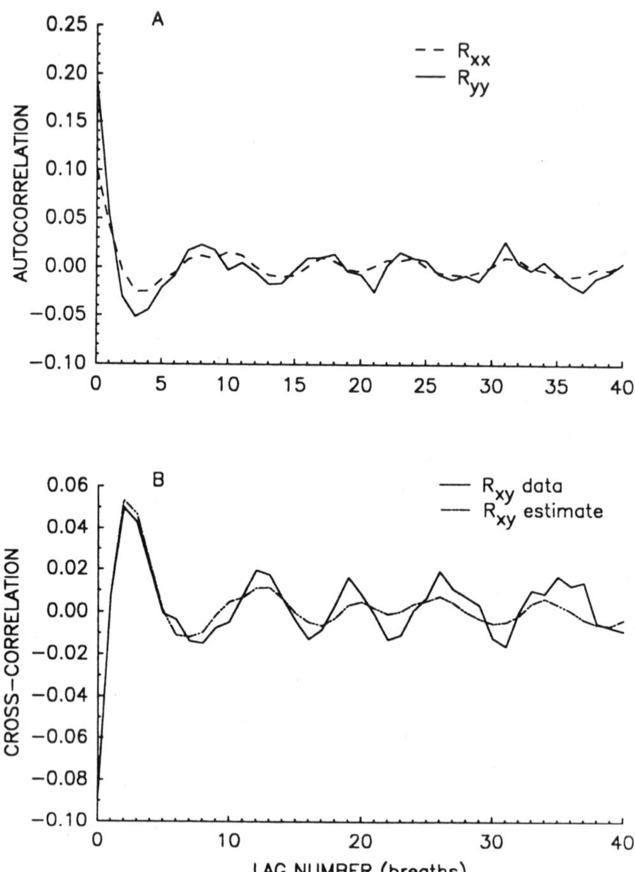

Fig. 7.　(A) Autocorrelations, R_{xx} (dashed curve) and R_{yy} (solid curve), computed from data shown in Fig. 6A. (B) Cross–correlation, R_{xy}, computed from data of Fig. 6A is compared with corresponding best–fit (broken line).

the estimates obtained from our procedure could turn out to be different from the corresponding values determined through the other tests simply because of a difference in CO_2 response slopes above and below eucapnia. In a recent study[24], we have pointed out other problems with one of these transient CO_2 inhalation techniques: the single–breath test.

The simulation study has shown that the broad–band spontaneous fluctuations in P_{ACO2} (or x) evoke responses from the central as well as the peripheral chemoreceptors. At the same time, however, there is insufficient power in the low frequencies of excitation to allow a clear delineation of the peripheral and central chemoreflex effects. The fact that we are unable to obtain unique parameter estimates from a double–exponential representation of the impulse response demonstrates this point. Consequently, the impulse responses derived with our technique contain influences from the central chemoreflexes that bias our estimates of peripheral gain. However, if central gain remains constant in a given subject, the bias also remains unchanged. Thus, we still have in this technique a means of quantifying <u>changes</u> in peripheral chemoreflex gain.

In our present analysis, the use of a breath–number time–scale implies an assumption of constant breath duration. This may not be justifiable as an approximation in some cases, particularly when significant pauses or short apneas occur frequently.

Another problem pertains to the optimum length of the data segment that one should employ in the computations. On the one hand, we are restricted by the assumption that the input and output time series should be stationary. On the basis of this consideration, it would be desirable to minimize the duration of the experiment. On the other hand, however, decreasing the number of data points leads to progressively greater inaccuracy and instability in the correlation estimates at higher lags.

A final limitation has been the assumption that the chemical component of ventilatory drive is dependent only on CO_2, and that hypoxia only modulates the sensitivity of the response to CO_2. While this assumption may be a very good approximation to reality during normoxia, it becomes much less valid during hypoxia when the physiologic perturbations in P_{aO2} may also affect ventilatory drive on a breath–to–breath basis. The poor fit between estimated R_{xy} and R_{xy} derived directly from the data in Fig. 7B is probably the result of excluding oscillations in P_{AO2} from our computations. Further work is needed to extend the estimation procedure to situations with two inputs (CO_2 and O_2) and one output (ventilatory drive).

This work was supported by NIH Grant RR–01861 and the Whitaker Foundation.

REFERENCES

1. P. Padget, The respiratory response to carbon dioxide, <u>Am. J. Physiol.</u> <u>83</u>:384–394 (1928).

2. D.J.C. Read, A clinical method for assessing the ventilatory response to carbon dioxide, <u>Australas. Ann. Med.</u> <u>16</u>:20–32 (1967).

3. D.J.C. Cunningham, P.A. Robbins and C.B. Wolff, Integration of respiratory responses to changes in alveolar partial pressures of CO_2 and O_2 and in arterial pH, <u>in</u>: "Handbook of Physiology – The Respiratory System II", A.P. Fishman, ed., Am. Physiol. Soc., Bethesda, MD (1987).

4. M.C.K. Khoo, R.E. Kronauer, K.P. Strohl and A.S. Slutsky, Factors inducing periodic breathing in humans: a general model, J. Appl. Physiol. 53:644–659 (1982).

5. S. Lahiri, C. Hsiao, R. Zhang, A. Mokashi and T. Nishino, Peripheral chemoreceptors in respiratory oscillations, J. Appl. Physiol. 58:1901–1908 (1985).

6. K.R. Chapman, E.N. Bruce, B. Gothe and N.S. Cherniack, Possible mechanisms of periodic breathing during sleep, J. Appl. Physiol. 64:1000–1008 (1988).

7. A. Berkenbosch, J. Heeringa, C.N. Olivier and E.W. Kruyt, Artificial perfusion of the ponto–medullary region in cats: A method for separation of central and peripheral effects of chemical stimulation of ventilation, Respir. Physiol. 37:347–364 (1979).

8. J.W. Bellville, B.J. Whipp, R.D. Kaufman, G.D. Swanson, K.A. Aqleh and D.M. Wiberg, Central and peripheral chemoreflex loop gain in normal and carotid body–resected subjects, J. Appl. Physiol. 46:843–853 (1979).

9. N.H. Edelman, P.E. Epstein, S. Lahiri and N.S. Cherniack, Ventilatory responses to transient hypoxia and hypercapnia in man, Respir. Physiol. 17:302–314 (1973).

10. P.A. McClean, E.A. Phillipson, D. Martinez and N. Zamel, Single breath of CO_2 as a clinical test of the peripheral chemoreflex, J. Appl. Physiol. 64:84–89 (1988).

11. S. Sohrab and S.M. Yamashiro, Pseudorandom testing of ventilatory response to inspired carbon dioxide in man, J. Appl. Physiol. 49:1000–1009 (1980).

12. P.A. Robbins, The ventilatory response of the human respiratory system to sine waves of alveolar carbon dioxide and hypoxia, J. Physiol. (London) 350:461–474 (1984).

13. S.M. Yamashiro and S. Ghazanshahi, Ventilatory CO_2 responses during hypocapnia in anesthetized cats, J. Appl. Physiol. 56:678–680 (1984).

14. P.J. Fleming, A.L. Goncalves, M.R. Levine and S. Woollard, The development of stability of respiration in human infants: Changes in ventilatory responses to spontaneous sighs, J. Physiol. (London) 347:1–16 (1984).

15. J.P. Cleave, M.R. Levine and P.J. Fleming, The control of ventilation: A theoretical analysis of the response to transient disturbances, J. Theor. Biol. 108:261–283 (1984).

16. M. Revow, S.J. England, H. O'Beirne and A.C. Bryan, A model of the maturation of respiratory control in the newborn infant, IEEE Trans. Biomed. Eng. 36:414–423 (1989).

17. M.C.K. Khoo and V.Z. Marmarelis, Estimation of peripheral chemoreflex gain from spontaneous sigh responses, Ann. Biomed. Eng. 17:557–570 (1989).

18. P.J. Brusil, "Statistical analysis of natural breathing patterns in supine man", (Ph.D. thesis) Harvard Univ., Cambridge, MA (1973).

19. L. Goodman, Oscillatory behavior of ventilation in resting man, <u>IEEE Trans. Biomed. Eng.</u> <u>11</u>:82–93 (1964).

20. J.A. Nelder and R. Mead, A simplex method for function minimization, <u>Computer Journal</u> <u>7</u>:308–313 (1965).

21. M.C.K. Khoo, Noninvasive tracking of peripheral ventilatory response to CO_2, <u>Int. J. Biomed. Comput.</u> (in press).

22. G.D. Swanson and J.W. Bellville, Step changes in end–tidal CO_2: Methods and implications, <u>J. Appl. Physiol.</u> <u>39</u>:377–385 (1975).

23. T. Chonan, Y. Kikuchi, W. Hida, C. Shindoh, H. Inoue, H. Sasaki and T. Takashima, Response to hypercapnia and exercise hyperpnea in graded anesthesia, <u>J. Appl. Physiol.</u> <u>57</u>:1796–1802 (1984).

24. M.C.K. Khoo, A model–based evaluation of the single–breath CO_2 ventilatory response test, <u>J. Appl. Physiol.</u> (in press).

Part III:

NEURAL CONTROL AND BREATHING PATTERN ANALYSIS

A THREE–PHASE MODEL OF RESPIRATORY RHYTHM GENERATION

Eugene N. Bruce

Dept. of Biomedical Engineering
Case Western Reserve University, Cleveland, OH

INTRODUCTION

A mathematical model of the central neural mechanisms of respiratory rhythm generation has been developed. This model is based substantially on a previously published qualitative model which assumes that the respiratory central pattern generator (RCPG) is formed of a neural network and that the respiratory cycle comprises three phases: inspiration, post–inspiration and expiration[1,2,3]. The objectives of evaluating quantitatively the behavior of this model are to determine whether a unique model with fixed parameters can quantitatively reproduce a range of behaviors attributed to the qualitative model, and to propose how afferent inputs from chemoreceptors and from slowly adapting pulmonary stretch receptors may be processed by the respiratory neurons in the brainstem based on model simulations.

MODEL STRUCTURE

Five respiratory neuronal groups are included: inspiratory, early inspiratory, post–inspiratory, late inspiratory, and expiratory neurons. Early–expiratory neurons are not included; within the model their functions may be assumed by post–inspiratory neurons, which have a similar firing pattern. Subcategories of post–inspiratory neurons are not distinguished in the model. The connections assumed between these groups (Fig. 1) have been proposed previously based on indirect evidence using various anatomical and electrophysiological techniques[4-7]. All connections are inhibitory except that from inspiratory to late–inspiratory neurons. In addition all groups have tonic inputs and all have self–excitatory feedback.

MODEL EQUATIONS

Each group of neurons is considered as a single unit whose output represents the average potential of the neurons in the group. The synaptic weights between the groups represent the average tendency of neurons of one group to project to and influence the potential of neurons of another group. The rate of change of the activity of one neural group is equal to a weighted sum of: (i) the activities of the other groups (each transformed by a sigmoidal function), (ii) a tonic input, (iii) a positive self–feedback, and (iv) a decay term proportional to the output of the group. This can be expressed mathematically[8] for each group X_i as:

$$\frac{dX_i}{dt} = -a_i \cdot X_i + \sum_{j=1,\neq i}^{5} W_{ij} \cdot S(X_j) + W_{ii} \cdot S(X_i) + T_i$$

where W_{ij} is the weighting of the connection from group "j" to group "i", S is a sigmoidal function[9] and T_i is the tonic input.

Because of the limited data available and the large number of parameters, we estimated the parameters of each group in isolation using the ideal patterns of activity of the different groups as diagrammed by von Euler[5]. To estimate the parameters of one neural group we assumed that all inputs from the other groups were the constructed ideal waveforms. The positive self–feedback parameters (i.e. the W_{ii}) were chosen arbitrarily to be 1.1 times the a_i parameters.

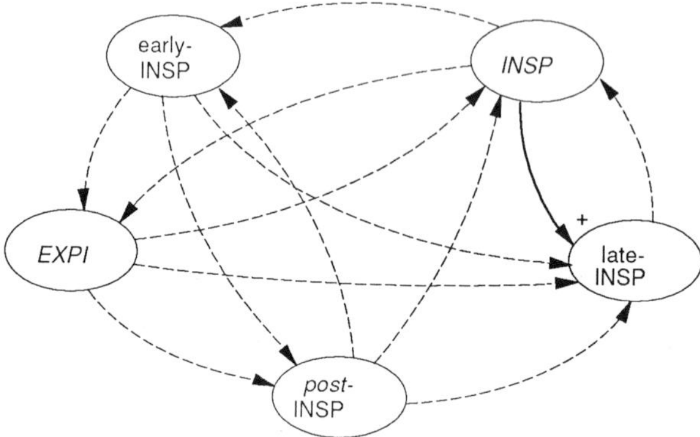

Fig. 1.   Diagram of the model. All connections are inhibitory except for that from INSP to late–INSP. Each group also receives a tonic input and positive self–excitation which are not shown.

Pulmonary stretch receptor afferent activity carried by the vagus nerves was simulated by passing the inspiratory group activity, transformed by the sigmoidal function S, through a simple RC lowpass filter. The time constant was chosen so that simulated stretch receptor activity would decay to baseline before the end of expiration.

SIMULATION RESULTS

The model produces a stable limit cycle and generally reproduces the features of the firing patterns of the 5 neuronal groups (Fig. 2). The early–inspiratory group, however, commences firing too soon in expiration. This behavior probably reflects the fact that in the model this neuronal group is the primary source of inhibition which terminates expiratory activity. Adding rebound excitation to this group permits it to develop sufficient inhibition to stop expiration with a later onset, but evidence for this is lacking. Alternatively, the sharp phase transition that is required could be produced by interaction of these modeled pools with a group of pacemaker neurons[10].

When simulated feedback from vagal pulmonary stretch receptors is made to excite late inspiratory neurons and inhibit early inspiratory neurons, the model quantitatively reproduces previous observations of the expiratory–prolonging effects of pulses and steps of vagal afferent activity presented in expiration (Fig. 3). In addition the model reproduces expected respiratory cycle timing and amplitude responses to change of chemical drive both in the absence and in the presence of simulated stretch receptor feedback (Fig. 4). Changes in chemical drive are simulated by scaling all tonic inputs, except that to the late–inspiratory pool, uniformly. The latter tonic input is changed by about 20% of the change to other tonic inputs in order to prevent alteration of cycle timing as chemical drive changes without vagal stretch receptor feedback.

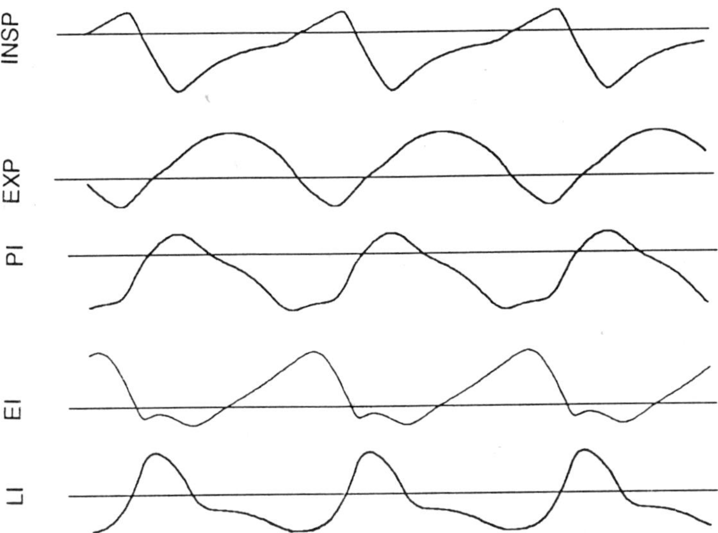

Fig. 2.   The 5 waveforms of the final model at relative
chemical drive equals 8, with simulated vagal
PSR afferent feedback. Cycle duration can be
scaled by scaling parameters.

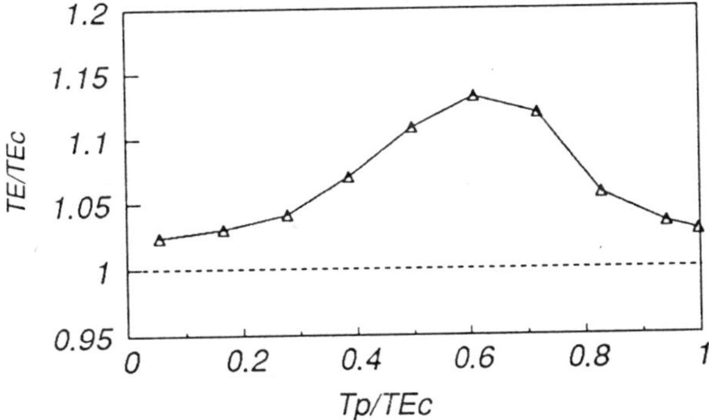

Fig. 3.   Expiratory duration (TE) relative to its control value (TEc) when a short (i.e., 5.6%
of TEc) pulse of simulated vagal pulmonary
stretch receptor activity is delivered at various times (Tp) in expiration. Vagal input
inhibits early–INSP neurons and excites late–
INSP neurons.

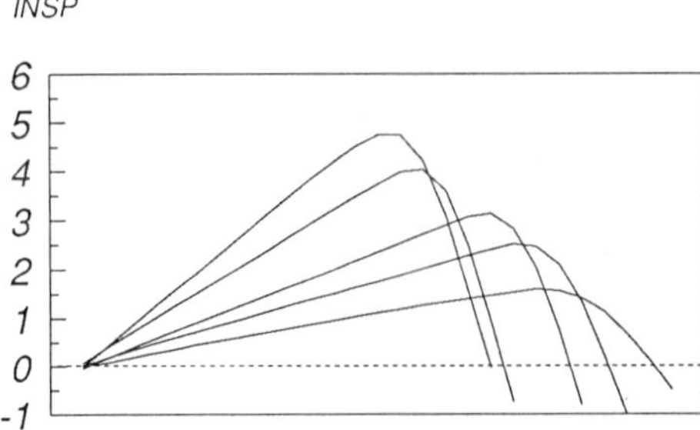

Fig. 4. Positive portions of inspiratory neuronal group waveform at various relative levels (i.e., 1.0, 1.5, 2, 3, 4) of chemical drive in the presence of simulated vagal stretch receptor feedback.

When brief pulses were added to vagal afferent activity at various times during the respiratory cycle, both weak and strong rhythm resetting responses were observed, with evidence for a singularity at the inspiratory–to–expiratory phase transition (Fig. 5). Similar responses have been noted experimentally in cats with superior laryngeal nerve stimulation[11]. In the model it is clear that the singularity occurs because the vagal stretch receptor feedback has two opposing actions: 1) it inhibits inspiration and indirectly acts to increase respiratory rate; 2) it prolongs expiration and acts to decrease rate. Near the phase transition small changes in the timing of the vagal afferent input cause large changes in the balance between these two effects.

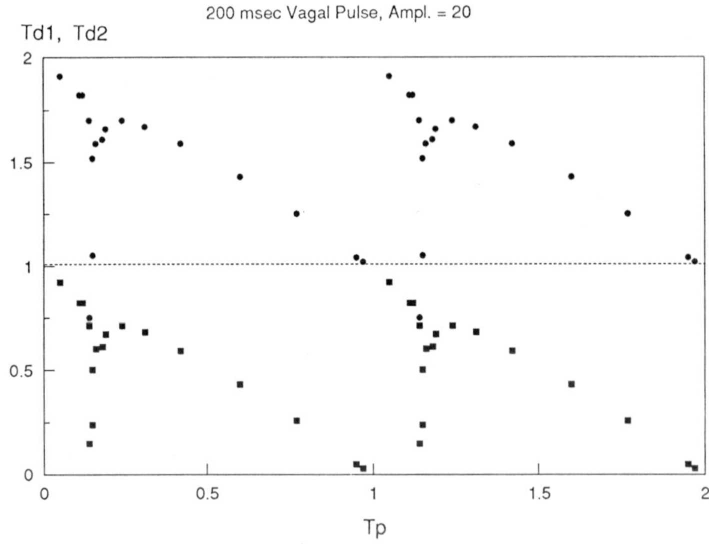

Fig. 5. Phase response curve resulting from a brief pulse of simulated pulmonary stretch receptor afferent activity, showing evidence of phase singularity. Tp is the time from beginning of inspiration to the pulse. Td1, Td2 – times from pulse to beginnings of first and second succeeding inspirations.

CONCLUSIONS

The present model has shown the feasibility of the qualitative concepts proposed by Richter and colleagues for generating a respiratory–like rhythm. Some details of the firing patterns of the neuronal groups are not reproduced well, however, and further refinement is necessary. Such a model can help to identify the critical experimental information which is necessary for this refinement. On the other hand the model can predict quantitatively the responses of respiratory pattern variables to moderate changes in chemical drive and vagal PSR afferent activity and has the possibility of extension to include other afferent inputs. Thus this model could be incorporated with chemical control models of respiration, which traditionally have either ignored respiratory rhythmogenesis or imposed algebraic relationships between pattern variables without regard to dynamic mechanisms.

REFERENCES

1. D. W. Richter, Generation and maintenance of the respiratory rhythm. J. Exp. Biol. 100:93–107 (1982).

2. D. W. Richter, D. Ballantyne, and J. E. Remmers, How is the respiratory rhythm generated? NIPS 1:109–112 (1986).

3. D. W. Richter, D. Ballantyne, and J.E. Remmers, The differential organization of medullary post–inspiratory activities, Pflug Arch. 410:420–427 (1987).

4. C. V. Euler, On the central pattern generator for the basic breathing rhythmicity, J. Appl. Physiol. 55:1647–1659 (1983).

5. C. V. Euler, Brainstem mechanisms for generation and control of breathing pattern. in: "Handbook of Physiology. The Respiratory System II," Amer. Physiological Soc, Washington, pp. 1–67 (1986).

6. J. L. Feldman, Neurophysiology of breathing in mammals. in: "Handbook of Physiology. The Nervous System IV," Amer. Physiological Society, Washington, pp. 463–524 (1986).

7. S. E. Long and J. Duffin, The medullary respiratory neurons: a review, Can. J. Physiol. Pharmacol. 62:161–182 (1984).

8. S. Geman and M. Miller, Computer simulation of brainstem respiratory activity, J. Appl. Physiol. 41:931–938 (1976).

9. J. A. H. Q. V. Dooren and A. Vis, A re–investigation of the Geman–Miller respiratory oscillator model, Biol. Cybern. 44:205–210 (1982).

10. J. Smith and J. L. Feldman, Central respiratory pattern generation studied in an in vitro mammaliam brainstem–spinal cord preparation, in: "Respiratory Muscles and their Neuromotor Control," Alan R. Liss, ed., pp. 27–36 (1987).

11. D. Paydafar, F. L. Eldridge, and J. P. Kiley, Resetting of mammalian respiratory rhythm: existence of a phase singularity, J. Appl. Physiol.: Reg. Integ. Comp. Physiol. 19:R721-R727 (1986).

CONCEPTUAL MODEL OF VENTILATORY MUSCLE RECRUITMENT AND DIAPHRAGMATIC FATIGUE

Gary C. Sieck

Department of Biomedical Engineering
University of Southern California
Los Angeles, California

INTRODUCTION

In the diaphragm as with other respiratory muscles, motor units (a motoneuron and the muscle fibers it innervates; Fig. 1) are the final effector pathway by which the central nervous system controls the level of ventilation. The forces generated by the diaphragm are controlled by changing the number of activated motor units (recruitment coding) and/or by modifying the discharge frequency of recruited units (frequency coding). As in other skeletal

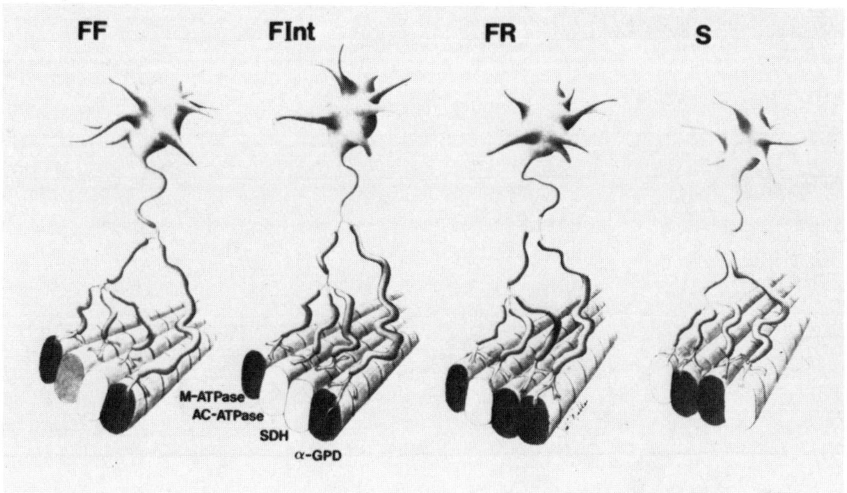

Fig. 1. Diaphragm motor units are comprised of phrenic motoneurons located in the cervical spinal cord and a set of muscle fibers (the muscle unit) with specific histochemical staining profiles. The muscle fiber types within each unit are distinguished by differences in staining for myofibrillar AT-Pase (after alkaline, M-ATPase and acid, AC-ATPase, preincubation), succinate dehydrogenase (SDH) and alpha-glycerophosphatase (GPD). From Sieck[3].

muscles, the motor units that comprise the diaphragm vary considerably in their contractile and fatigue properties[1]. The overall strength of the diaphragm and its susceptibility to fatigue depend upon the motor units recruited by the central nervous system. For example, under conditions requiring prolonged low level force production by the diaphragm (e.g., quiet breathing), the nervous system might recruit only those units that are fatigue resistant. In contrast, during other conditions where the diaphragm generates increased levels of force (e.g., sneezing, gagging), the recruitment of more fatigable motor units would also be required. Thus, it is important to consider what types of diaphragm motor units are available for recruitment, both in the generation of a respiratory rhythm and in the production of nonventilatory forces.

MOTOR UNIT CLASSIFICATION

Burke et al.[2] established standard criteria for the classification of motor units. In their scheme, fast-twitch units were distinguished from slow-twitch units by the presence of sag in unfused tetani (usually at stimulation rates between 5 and 20 pulses per sec (pps; Fig. 2). In most muscles, including the diaphragm[1], the presence or absence of sag generally correlates with twitch contraction time (CT, time to peak tension; Fig. 3). Burke et al.[2] further subclassified fast-twitch units into three types based on difference in fatigue resistance (Fig. 2), i.e., fast-twitch fatigable (FF), fast-twitch fatigue intermediate (FInt), and fast-twitch fatigue resistant (FR). This classification scheme permits comparison of motor unit properties across different muscles and species, as well as quantification of muscle adaptations to altered use (e.g., in terms of varying motor unit proportions).

Several studies have suggested that both recruitment and frequency coding of diaphragm motor units are important in the neural control of diaphragm force generation[4,5,6]. Yet, the mechanical correlates of this neural control of diaphragm motor unit activation was only recently established[1]. The purpose of the present study was to provide a conceptual model describing how the forces generated by the diaphragm during different ventilatory and non-ventilatory behaviors might be achieved by varous combinations of motor unit recruitment and frequency coding.

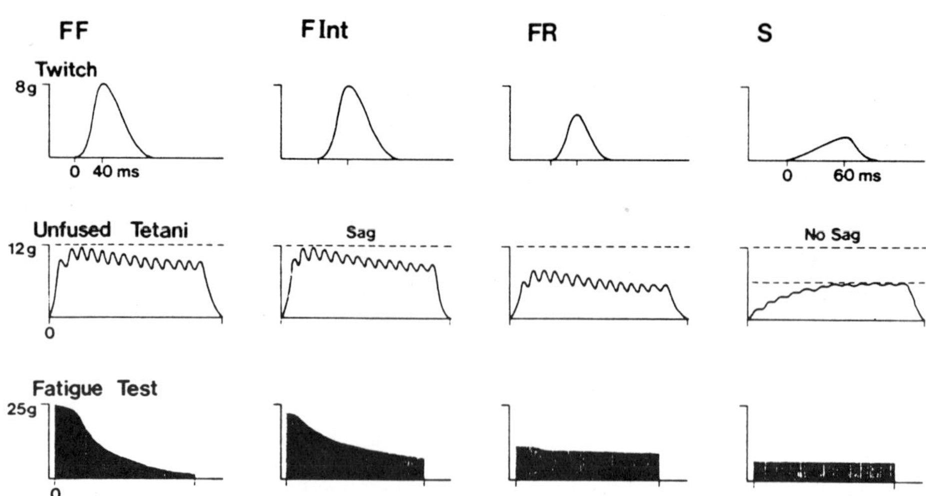

Fig. 2. Summary of the physiological criteria established by Burke et al.[2] for the classification of different motor unit types. From Sieck[3].

114

MEASUREMENT OF DIAPHRAGM MOTOR UNIT CONTRACTILE PROPERTIES

Figure 4 summarizes the procedures we used in a recent study[1] to characterize the contractile and fatigue properties of motor units in the cat diaphragm. During the experiments, animals were anesthetized with pentobarbital sodium (35 mg/kg, ip). Arterial blood pressure was monitored and maintained by intravenous infusion of lactated Ringers and Hetastarch. Core body temperature was maintained at 37°C using radiant heat. The trachea was intubated, and end-tidal CO_2 was monitored. The level of ventilation was also assessed by measuring arterial blood gases.

Animals were positioned in a stereotaxic frame with their head and vertebral column rigidly fixed. The diaphragm was exposed and pairs of fine wire electrodes were inserted into the costal (ventral, middle and dorsal) and crural regions of the muscle. Electromyographic (EMG) signals were filtered and amplified. The central tendon of the diaphragm was clamped near the insertion of muscle fibers in the sternocostal region, and the clamp was rigidly secured to the stereotaxic frame to provide a fixed reference point for isometric tension measurements. Care was taken during this procedure to avoid occluding any vasculature in the central tendon. The origin of fibers along the costal margin was detached from the rest of the rib cage by cutting transversely through ribs 9 to 13 and by sectioning the point of convergence of the fused ribs to the sternum. The freed costal margin was attached in series to a force transducer, and the length of muscle fibers was adjusted by pulling outward at the costal margin. The

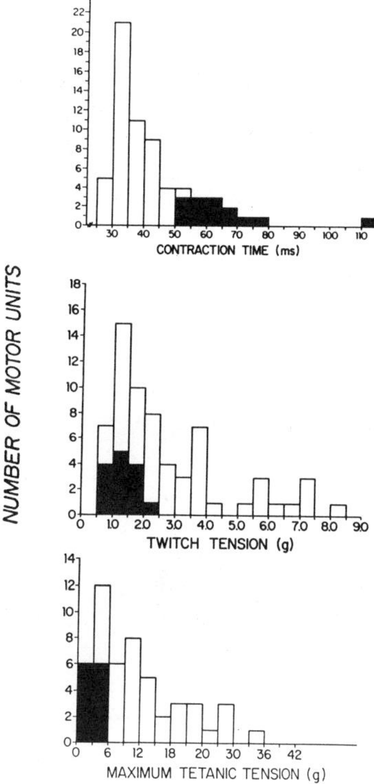

Fig. 3. Distribution of diaphragm motor unit isometric twitch contraction times (CT), peak twitch tension (Pt) and maximum tetanic tensions (Po). Slow-twitch units (i.e., no sag) are indicated by the filled bars and fast-twitch units by the open bars. From Fournier and Sieck[1].

temperature of the exposed muscle was maintained at 35°C using radiant heat. The diaphragm was also periodically coated with warm mineral oil to prevent drying.

The cervical spinal cord from C_2 to C_7 was exposed by a dorsal lamenectomy. The spinal cord was then transected at C_3 to block descending rhythmic drive to the diaphragm and intercostal muscles. Thereafter, the animal was mechanically ventilated to maintain end-tidal CO_2 at 4%. Spinal cord transection at C_3 resulted in a precipitous drop in arterial pressure. Experiments were not continued until mean arterial pressure was stabilized above 80 mm Hg.

In previous studies[8,9], we demonstrated somatotopy in the segmental innervation of the cat diaphragm with C_5 primarily innervating the ventral portions of the muscle and C_6 innervating the more dorsal portions (Fig. 5). We also found that fibers belonging to single motor units innervated by C_5 axons were not distributed throughout the diaphragm, but instead were localized to a relatively small region of the muscle[9]. This restricted territory of motor unit fibers was important technically because it permitted us to accurately measure the contractile forces of diaphragm units. Diaphragm fibers are radially oriented. If unit fibers were scattered across a wide territory of the muscle, the forces produced by the dispersed fibers would tend to cancel one another. Therefore, in our study1, contractile forces were measured only for motor units with muscle fibers located in the right sternocostal region of the diaphragm. The restricted location of unit fibers was confirmed by the presence of evoked EMG responses in the sternocostal region and by the absence of evoked responses in other diaphragm regions. In recent studies using the technique of glycogen depletion, we have also verified the restricted distribution of diaphragm motor unit fibers.[10]

Units were isolated by microdissection and stimulation (0.2 ms rectangular pulses) of C_5 ventral root filaments. The isolation of a single unit was verified by the presence of all-or-none evoked EMG and twitch force responses across a 5-fold range of stimulus intensities above threshold. Thereafter, stimulus intensity was set at 1.5 times threshold (i.e., supramaximal), and the length of muscle fibers was adjusted until maximal twitch force

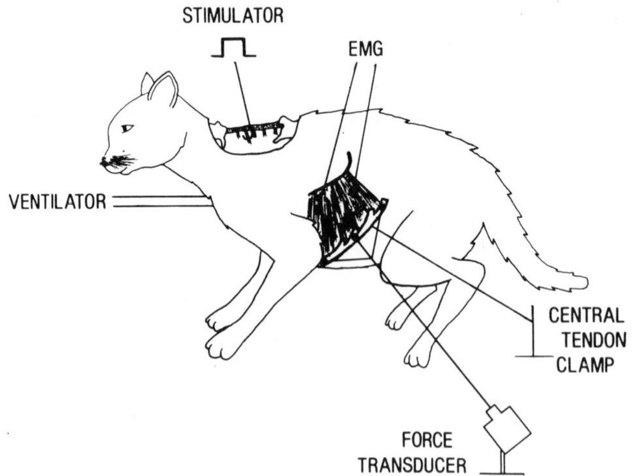

Fig. 4. Summary of the procedures used to isolate and characterize single motor units in the cat diaphragm. From Sieck[7].

responses were obtained. This corresponded to a preload tension of 75 to 100 g. Units were activated using a computer-controlled protocol for stimulation at a range of frequencies from 1 to 100 pps. Twitch CT and peak twitch tension (Pt) were measured, and maximum tetanic tension (Po) was determined. The dependency of force production on stimulus frequency was characterized by constructing force/frequency curves (Fig. 6).

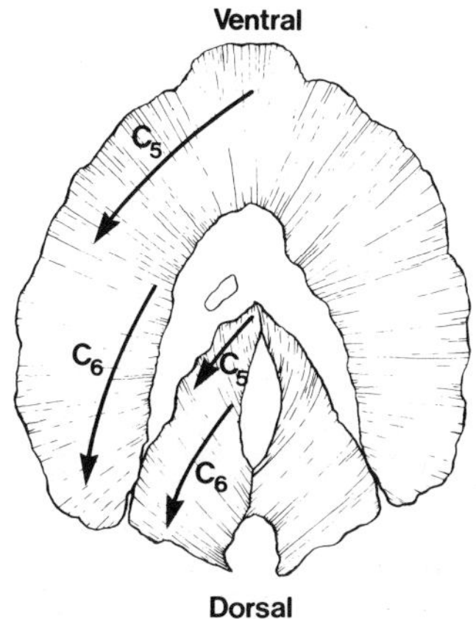

Fig. 5. Summary of the segmental innervation of the cat
diaphragm. From Fournier and Sieck[8].

The fatigue resistance of units was assessed using a standard fatigue test[2] which employed repetitive stimulation at 40 pps in trains of 330 ms duration with 1 train repeated each sec. A fatigue index (fi) was calculated as the ratio of tension produced after 2 min of stimulation to the initial tension. Motor units were classified as: Fast-twitch fatigable (FF, fi<0.25); Fast-twitch fatigue intermediate (FInt, 0.25<fi<0.75); Fast-twitch fatigue resistant (FR, fi>0.75); or Slow-twitch (S, fi>0.75), based on the criteria established by Burke et al.[2]

In the diaphragms of 10 cats, 53 units were completely characterized, and another 15 were partially analyzed. We found that 41% of all units were type FF, 25% FInt, 4% FR, and 30% S. Figure 7 summarizes the differences in Pt and Po for these different motor unit types. It should be noted that at all stimulus frequencies, type S units generated significantly

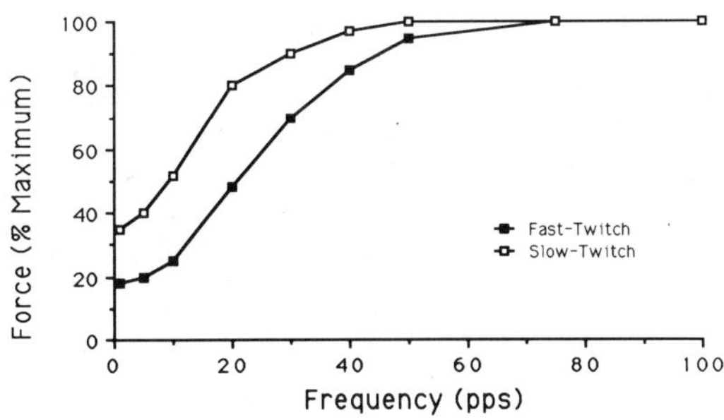

Fig. 6. The force generated by diaphragm motor units
depended on the frequency of stimulation. Average relative force responses of 4 slow- and 5
fast-twitch units from a single diaphragm. From
Sieck[7].

lower tensions (P<0.01) than the different fast-twitch unit types.

Because of these differences in tension, the relative contribution of each unit type to the total force generating capacity of the diaphragm is not reflected by their proportions within the muscle. Although 30% of all units in the diaphragm were type S, these units were estimated to contribute only 11% of the total maximum tetanic tension of the muscle (Fig. 8). Conversely, FF units comprised only 41% of all units in the diaphragm, yet were estimated to contribute 55% to the total force generating capacity of the muscle. Due to differences in force/frequency responses (Fig. 6), type S units contributed more to total muscle force at lower rates of activation than at higher activation rates. The opposite was true for fast-twitch units. For example, between 10 to 20 pps, type S units contributed 20 to 23% to total muscle force compared to 11% at 50 pps and above. In contrast, between 10 to 20 pps, FF units contributed 47 to 49% of total force compared to 55% at activation rates above 50 pps.

SPONTANEOUS DISCHARGE RATES OF DIAPHRAGM MOTOR UNITS

In a previous study[12], we examined the spontaneous discharge rates of diaphragm motor units in unanesthetized cats during different sleep-waking states. Fine wire electrodes were inserted into the costal or crural regions of the diaphragm. The animals were also instrumented with electrodes for monitoring sleep-waking state (e.g., cortical EEG, electro-occulogram, depth electrodes in the lateral geniculate nucleus, and EKG electrodes). The electrodes were channeled to the top of each animal's head and connected to a plug which was then cemented to the calvarium. One week after recovery from this initial surgery, an external cable was attached to this head plug and the various electrical signals were amplified and appropriately filtered. The diaphragm EMG signals were band-pass filtered between 20 Hz and 3 KHz. Single motor unit discharge was discriminated based on constant waveform shape and amplitude.

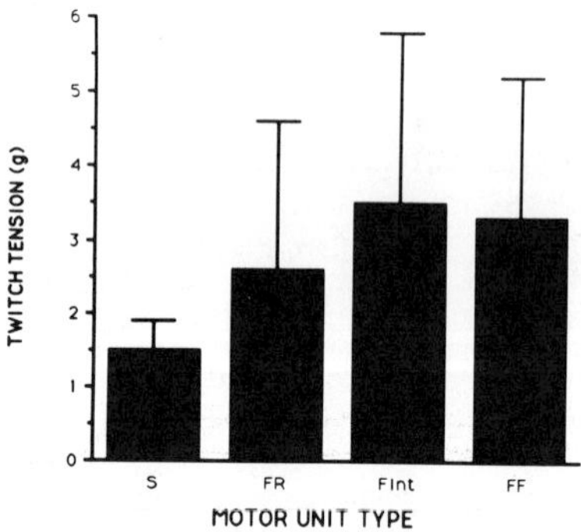

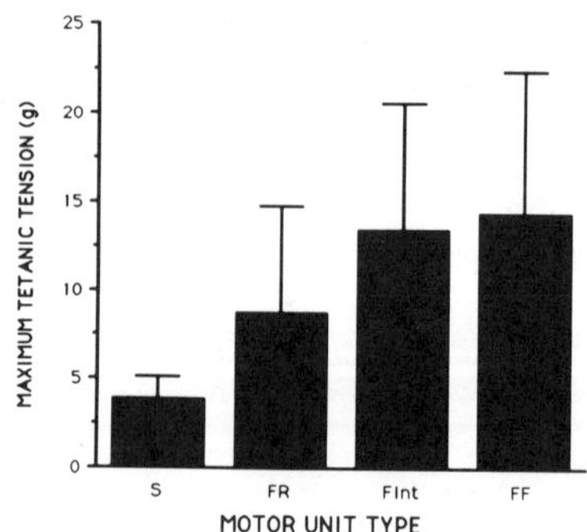

Fig. 7. Mean (± standard deviation) twitch (Pt) and maximum tetanic (Po) tensions generated by each unit type in the diaphragm. From Sieck[7].

Unit discharge patterns were assessed using a point process autocorrelation analysis. Most of the diaphragm motor units that were consistently recruited during quiet breathing showed strong peaks in the autocorrelations of their discharge (Fig. 9B&C). Motor units that were recruited early during the inspiratory period had strong autocorrelation peaks indicating modal discharge rates ranging from 8 to 12 pps (Fig. 9C). Motor units that were also consistently recruited but later during the inspiratory period, also had strong autocorrelation peaks, indicating modal discharge rates ranging from 13 to 16 pps (Fig. 9B). These very strong autocorrelation peaks in diaphragm motor unit discharge suggested that little variation in discharge rate occurred during quiet breathing. Thus, changes in diaphragm force during eupnea would appear to be primarily accomplished by recruitment rather than frequency coding. However, at times during quiet breathing, some diaphragm motor units were recruited very late during inspiration. These "late" units had modal discharge rates ranging from 20 to 25 pps, but their autocorrelation peaks were typically very weak, indicating considerable variation in discharge rate. It is possible that for these late diaphragm motor units frequency coding is more important in the control of force generation. Unfortunately, during ventilatory efforts requiring additional force generation by the diaphragm (e.g., increased chemical drive), it was not possible to discriminate single unit discharge because of the recruitment of additional motor units.

Table 1. Transdiaphragmatic pressures (Pdi) generated during different ventilatory and non-ventilatory behaviors. The Pdi responses were normalized for the maximum Pdi generated by bilateral phrenic nerve stimulation (Pdi max). From Sieck and Fournier[11].

Cat #	Pdi max (cm H_2O)	Eupnea	Chem Drive	Trach Occl	Gag Refl	Sneeze
1	37	11%	34%	64%	121%	
2	33	15%	36%	42%	109%	115%
3	37	10%	27%	47%	97%	
4	31	11%	17%	37%	97%	106%
5	34	13%	25%	56%	111%	
Avg	34.4	12.0%	27.8%	49.2%	107.0%	110.5%
±SD	±2.6	±2.0	±7.6	±10.6	±10.2	±6.5

FORCES GENERATED BY THE DIAPHRAGM DURING DIFFERENT VENTILATORY AND NON-VENTILATORY BEHAVIORS

The forces generated by the cat diaphragm were estimated by measuring transdiaphragmatic pressure (Pdi) using a double lumen gastro-esophageal catheter[11]. The distal balloon of the catheter was positioned in the stomach to measure gastric pressure (Pga), and the proximal balloon was positioned in the esophagus just above the diaphragm to record pleural pressure (Ppl). The two lines of the catheter were connected to differential pressure transducers (Statham Model P45). Pga and Ppl were monitored separately, and Pdi was then calculated as the difference between them. Animals were studied while anesthetized: induced by intramuscular injection of ketamine (35mg/kg) and xylazine (5mg/kg) and maintained by intravenous injection of alpha-chloralose (40 mg/kg) and urethane (250 mg/kg). Arterial blood pressure was monitored throughout the experiment and maintained by intravenous infusion of lactated Ringers. The trachea was intubated and ventilatory level was assessed by monitoring end-tidal CO_2. Animals were studied both in the prone and supine position with the abdominal wall bound. Pdi responses were determined while the trachea was occluded.

To determine maximum Pdi, the phrenic nerves were isolated in the neck on both sides and stimulated supramaximally (0.2 ms duration rectangular pulses). The Pdi responses to stimulus frequencies ranging from 1 to 100 pps were measured. Maximum Pdi (Pdi max) responses were consistently achieved at a stimulus rate of 75 pps. The Pdi measured during different ventilatory and non-ventilatory behaviors were normalized for Pdi max.

Table 1 summarizes the Pdi responses measured during different ventilatory and non-ventilatory behaviors in 5 adult cats. Each measurement was repeated at least 2 times and in some cases (Pdi max, eupnea, tracheal occlusion) as many as 6 replicate measurements were obtained at various times throughout the experiment. Replicate measurements showed very little variation (10% or less). The Pdi max generated by bilateral phrenic nerve stimulation was very consistent across animals, ranging from 31 to 37 cm H_2O.

The Pdi response generated by single pulse stimulation of the phrenic nerves (i.e., twitch Pdi) was generally about 25% of Pdi max (i.e., average 8.2±1.8 cm H_2O). In contrast, the Pdi generated during eupnea ranged from 10 to 15% of Pdi max. This clearly demonstrated that during quiet breathing, only a fraction of the diaphragm motor unit pool is recruited. When the chemical drive for breathing was increased by having the animals inspire a gas mixture of 5% CO_2 and 10% O_2, the maximum steady state Pdi response was achieved within 90 to 120 sec and averaged 27.8% of Pdi max. Thus, even during increased chemical drive, it is unlikely that all diaphragm motor units are recruited. In response to maintained airway occlusion, the force generated by the diaphragm increased to approximately 49% of Pdi max (usually after the fourth or fifth inspiratory effort). Clearly, to generate this level of force additional motor units were recruited. It is also likely that unit discharge rates also increased. The Pdi generated during the gag reflex, elicited by mechanical stimulation of the oropharynx, ranged from 97 to 121% of Pdi max. In two animals, Pdi was also measured during sneezing behaviors. As in the case of the gag reflex, the Pdi generated during sneezing exceeded the Pdi max produced by bilateral phrenic nerve stimulation. Undoubtably during these expulsive, nonventilatory behaviors of the diaphragm, there is full recruitment of the motor unit pool and maximum discharge rates.

ESTIMATION OF DIAPHRAGM MOTOR UNIT RECRUITMENT

It has been shown that during most motor behaviors, motor units are recruited in an orderly fashion[13,14]. This recruitment order appears to depend mainly on the intrinsic

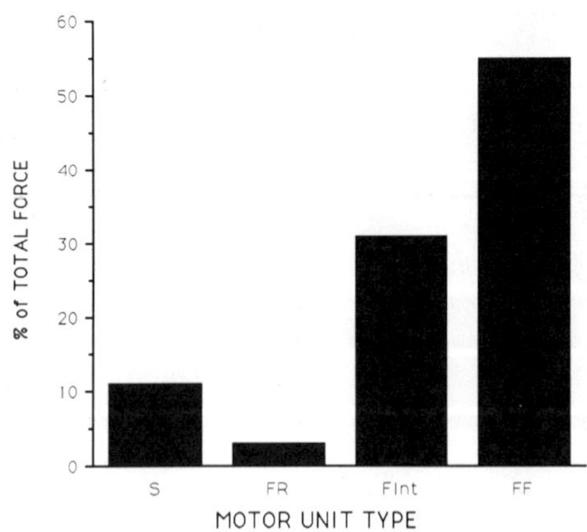

Fig. 8. Estimated relative contributions of each motor unit type to the maximum force generating capacity of the entire diaphragm. From Sieck and Fournier[11].

properties of motoneurons[14,15,16]. Smaller motoneurons with higher membrane resistance and slower axonal conduction velocities are recruited first, whereas larger motoneurons with lower membrane resistance and faster axonal conduction velocities are recruited later. Recruitment order also correlates with motor unit type, since smaller motoneurons generally innervate type S motor units and larger motoneurons innervate fast-twitch units[13,15,16]. Further, it has been demonstrated that there is a direct correlation between motoneuron size and the tetanic tensions generated by motor units[13,15,16]. Thus, smaller motoneurons which are recruited first, innervate units that produce lower tetanic tensions and larger motoneurons which are recruited later, innervate units that produce greater tetanic tensions.

Recent studies[4,17] have suggested that the order of diaphragm motor unit recruitment during inspiration depends on the intrinsic properties of phrenic motoneurons. In the present study, we assumed that diaphragm units are recruited according to their fatigue resistance (i.e., type S units recruited first, followed in order by FR, FInt and FF units), and that all units of a specific type are maximally activated (i.e., maximum tetanic tension) before the next type is recruited (Fig. 10). Based on these assumptions, we conclude that the diaphragm forces necessary for normal quiet ventilation could be achieved by the recruitment of only fatigue resistant motor units. During more forceful ventilatory efforts (e.g., with increased chemical drive or against increased airway resistance), the recruitment of type FInt units would be required. As these units do show some fatigue, when repetitively activated (with a duty cycle of 0.33 as in the present study), the diaphragm would be susceptible to fatigue under such ventilatory conditions. Based on our model, non-ventilatory behaviors which require short bursts of maximal force (e.g., gagging, sneezing) would require maximal activation of all diaphragm motor units. Such high forces could not be sustained for any appreciable length of time without substantial fatigue. Using a similar model for motor unit recruitment in the medial gastrocnemius muscle of the cat, Walmsley et al.[18] also reported that most motor behaviors could be accomplished by the recruitment of fatigue resistant

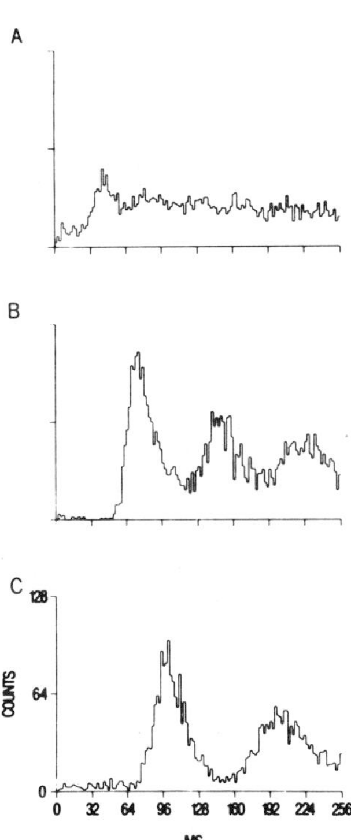

Fig. 9. Autocorrelation histograms of the discharge of diaphragm motor units. From Sieck et al.[12]

units. As in the diaphragm, these authors concluded that the recruitment of FF units would be necessary only during motor behaviors that required short duration bursts of high force output (e.g., jumping).

It should be emphasized again that our conceptual model for diaphragm motor unit recruitment assumes maximal activation (Fig. 10). However, at the spontaneous discharge rates that we observed (Fig. 9), type S units would generate only about 50 to 70% of their maximal force and type F units 25 to 50% (Fig. 6). At these discharge rates, the diaphragmatic forces generated during quiet breathing would require the recruitment of all fatigue resistant motor units (i.e., type S and FR) and possibly some FInt units. The diaphragm forces generated during increased ventilatory drive and during the expulsive, nonventilatory behaviors would require not only recruitment of additional motor units, but also an increase in the discharge rates of active motor units.

In summary, we conclude that quiet ventilatory forces can be generated by the progressive recruitment of fatigue resistant motor units in the diaphragm. During more forceful ventilatory and non-ventilatory efforts, both recruitment and frequency coding become important in the generation of diaphragm forces. However, with the recruitment of FInt and FF units during these more forceful behaviors, the diaphragm becomes more susceptibile to fatigue. The nervous system might adjust for this increased susceptibility to fatigue by limiting the duration of these more forceful ventilatory and nonventilatory behaviors, and/or by alternating diaphragm activation with the activation of synergist inspiratory muscles.

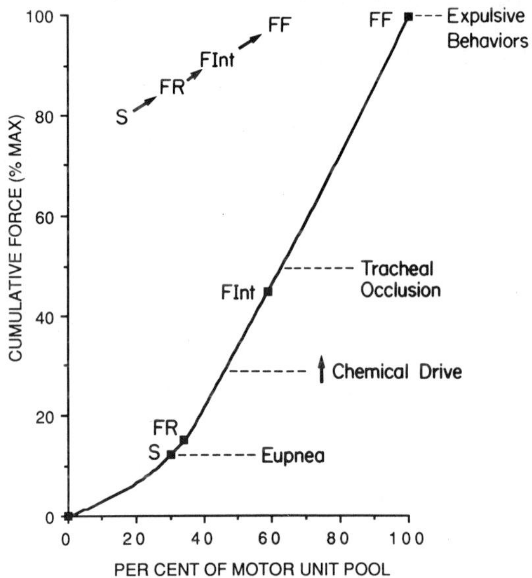

Fig. 10. The cumulative force (% of maximum tetanic tension) generated by the progressive recruitment of diaphragm motor units. From Sieck and Fournier[11].

ACKNOWLEDGEMENTS

I am greatful for the valuable contributions of Dr. Mario Fournier in these studies and for the assistance of Mr. Weizheng Wang in the preparation of this manuscript. This research was supported by grants from the National Heart, Lung, and Blood Institute (HL34817 and HL37680) and from the American Lung Association of Los Angeles County.

REFERENCES

1. M. Fournier, and G.C. Sieck, Mechanical properties of muscle units in the cat diaphragm. J. Neurophysiol. 59:1055-1066 (1988).
2. R.E. Burke, P.T. Levine, and F.E. Zajac III, Physiological types and histochemical profiles in motor units of the cat gastrocnemius. J. Physiol. Lond. 234:723-748 (1973).
3. G.C. Sieck, Diaphragm muscle: Structural and functional organization, in: "Respiratory Muscles: Function in Health and Disease," M.J. Belman, ed., Clinics in Chest Medicine 9:195-209 (1988).
4. T.E., Dick, F.J. Kong, and A.J. Berger, Correlation of recruitment order with axonal conduction velocity for supraspinally driven diaphragmatic motor units. J. Neurophysiol. 57:245-259 (1987).
5. G. Hilaire, P. Gauthier, and R. Monteau. Central respiratory drive and recruitment order of phrenic and inspiratory laryngeal motoneurones. Respiration Physiol. 51:341-359 (1983).
6. S. Iscoe, J. Dankoff, R. Migicovsky, and C. Polosa. Recruitment and discharge frequency of phrenic motoneurones during inspiration. Respiration Physiol. 26:113-128 (1976).
7. G.C. Sieck, Recruitment and frequency coding of diaphragm motor units during ventilatory and non-ventilatory behaviors, in: "Respiratory Control: Modelling Perspective," G.D. Swanson and F.S. Grodins, eds., Plenum, New York, NY, (1989).
8. M. Fournier, and G.C. Sieck. Somatotopy in the segmental innervation of the cat diaphragm. J. Appl. Physiol. 64:291-298 (1988).
9. M. Fournier, and G.C. Sieck. Topographical projections of phrenic motoneurons and motor unit territories in the cat diaphragm, in: "Respiratory Muscles and Their Neuromotor Control," G.C. Sieck, S.C. Gandevia, and W.E. Cameron, eds., Alan R. Liss, New York, NY, pp. 215-226 (1987).
10. G.C. Sieck, M. Fournier, and J.G. Enad. Fiber type composition of muscle units in the cat diaphragm. Neurosci. Lett. 97: 29-34 (1989).
11. G.C. Sieck, and M. Fournier, Diaphragm motor unit recruitment during ventilatory and non-ventilatory behaviors. J. Appl. Physiol. 66:2539-2545 (1989).
12. G.C. Sieck, R.B. Trelease, and R.M. Harper. Sleep influences on diaphragmatic motor unit discharge. Exp. Neurol. 85:316-335 (1984).
13. R.E. Burke, Motor units: Anatomy, physiology, and functional organization, in: "Handbook of Physiology, The Nervous System, Motor Control," J.M. Brookhart and V.B. Mountcastle, eds., Am. Physiol. Soc., Bethesda, MD, Vol II, Part 1, sect. 1, pp. 345-422 (1981).
14. E. Henneman, and L.M. Mendell. Functional organization of motoneuron pool and its input, in: "Handbook of Physiology, The Nervous System, Motor Control," J.M. Brookhart and V.B. Mountcastle, eds., Am. Physiol. Soc., Bethesda, MD, Vol. II, Part 1, Sec. 1, pp. 423-507 (1981).
15. J.W. Fleshman, J.B. Munson, G.W. Sypert, and W.A. Friedman. Rheobase, input resistance, and motor unit type in medial gastrocnemius motoneurons in the cat. J. Neurophysiol. 46:1326-1338 (1981).
16. F.E. Zajac, and J.S. Faden. Relationship among recruitment order, axonal conduction velocity, and muscle unit properties of type-identified motor units in cat plantaris muscle. J. Neurophysiol. 53:1303-1322 (1985).
17. J.S. Jodkowski, F. Viana, T.E. Dick, and A.J. Berger. Electrical properties of phrenic motoneurons in the cat: correlation with inspiratory drive. J. Neurophysiol. 58:105-124 (1987).
18. B. Walmsley, J.A. Hodgson, and R.E. Burke. Forces produced by medial gastrocnemius and soleus muscles during locomotion in freely moving cats. J. Neurophysiol. 41:1203-1216 (1978).

PHASE RESETTING OF THE RESPIRATORY OSCILLATOR–
EXPERIMENTS AND MODELS

Frederic L. Eldridge

Depts. of Medicine and Physiology, University of North Carolina
Chapel Hill, North Carolina 27599

Breathing involves an oscillatory process that causes rhythmic alternation between inspiration and expiration. Either of these parts of the respiratory cycle can be perturbed by various stimuli. When the oscillator recovers its initial rhythm after such stimuli, its phase may be reset relative to that of the control. The amount of resetting depends upon two factors, one of which is the strength of the stimulus and the other the time in the cycle at which it is given.

The basic experimental paradigm for experiments to study resetting is to present a brief stimulus to an ongoing rhythm. The effect of this stimulus on subsequent repeats of the rhythmic event is determined as a function of the phase of the cycle at which the stimulus was delivered. From the data generated, a topological resetting plot can be made of the relationship. The experiments are then repeated several times with stimuli of various strengths, ranging from very weak to very strong.

Important definitions[1] for understanding the resetting patterns include: old phase (ϕ), which is defined as the time from onset of a cycle, in our case the onset of inspiration, to the onset of a stimulus; and cophase (θ), which is the time from the end of the stimulus to the onset of the rescheduled breaths. The data are normalized by assigning a value of one to the average period of control breaths preceding the stimulus. The topological pattern of phase resetting is determined by constructing a plot of cophase vs. old phase for all stimuli of the same strength.

Using electrical stimulation of the mesencephalic reticular formation in the cat, we found continuous phase resetting patterns[2]: Type 1 for weak stimuli [Fig. 1, panels I (10 Hz) and II (20 Hz)] as defined in the figure; Type 0 resetting for strong stimuli [panels IV (30 Hz) and V (100 Hz)]; and unpredictable or ambiguous resetting [panel III (25 Hz)] when the intermediate strength stimulus was given at an old phase of approximately 0.4 of the cycle, at the transition from inspiration to expiration, i.e., a phase singularity. A three–dimensional plot made from the resetting patterns generated by the various strength stimuli depicted a helicoid surface whose axis represents the oscillator's phase singularity (see Fig. 4).

Modeling and Parameter Estimation in Respiratory Control
Edited by M.C.K. Khoo
Plenum Press, New York

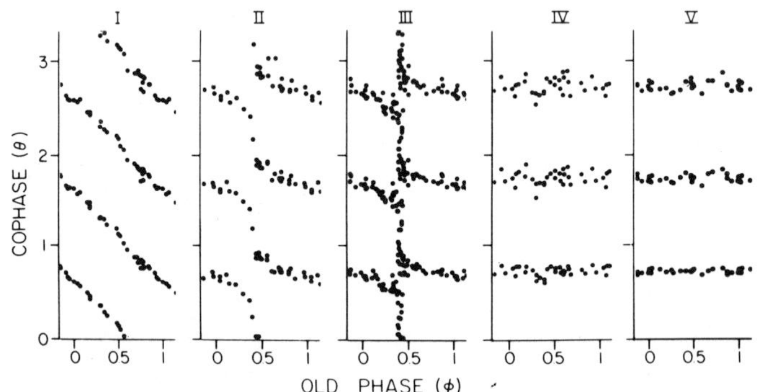

Fig. 1.   Definitions of phase resetting patterns; plots of resetting of
respiratory rhythms in cat with increasing strengths of midbrain stimulus, from weak (panel I, Type 1 resetting) to
strong (panel V, Type 0 resetting). Panel III shows ambiguous resetting with intermediate strength stimuli at old phase
of 0.4. (Reprinted from Ref. 2)

Similar findings occurred with a stimulus with an inhibitory effect, generated by electrical stimulation of the superior laryngeal nerve, except that the singularity occurred at the
transition from expiration to inspiration[3].

These findings have suggested that respiratory rhythm is generated by a continuous
limit–cycle oscillator, showing both type 0 resetting patterns with strong stimuli and a phase
singularity[1], and not by discrete on–switch, off–switch mechanisms which would show neither
of these features[4].

INFLUENCE OF STOCHASTIC PROCESSES

In contrast to the continuity noted above, in some animals we found resetting plots that
appeared to be discontinuous, particular at the intermediate stimulus strength and at the
old phase that should yield a singular response. An example is shown in panel III of Fig. 2.
Typical Type 1 resetting patterns occurred with weak stimuli (panels I and II), and Type 0
patterns with strong stimuli (panels IV and V). With intermediate strength stimuli (panel
III), the resetting pattern showed apparent discontinuities with clustering of cophases around
certain values despite a number of stimulus trials at the same old phase of approximately 0.4
of the cycle.

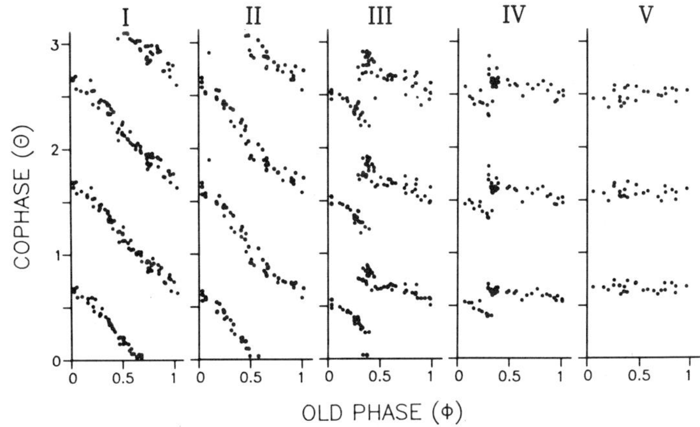

Fig. 2.   Plots of resetting of respiratory rhythm in cat with midbrain stimuli: weak stimuli (Type 1 resetting pattern) in plots I and II at left, strong (Type 0) in plots IV and V at right. Intermediate stimulus yields ambiguous pattern with apparent discontinuities of cophases.

We postulated that such apparent discontinuities, rather than contradicting the limit–cycle hypothesis, might be due to stochastic processes ("noise") inherent in the biological preparation. We therefore studied the resetting of rhythm in a model of a limit–cycle oscillator, and then the effect on that resetting of adding random "noise." For the modeling, we used a differential equation for a multivibrator that represents an example of relaxation oscillation, the Van der Pol limit–cycle oscillator[5], where

$$dy/dt \;=\; \varepsilon \cdot (1 - x^2) \cdot y - x \qquad \text{and}$$
$$y \;=\; dx/dt$$

Parameter ε is the model's coefficient of relaxation, which defines the shape of the limit cycle and the rate of the oscillator's return to the limit cycle after perturbation[5]. We chose a value of 2.0 for the model because it produces a fairly rapid ($\sim$ 0.5 of a cycle) return to the limit cycle after a perturbation. The equation was incorporated into a digital computer FORTRAN program that generated the limit cycles. The protocol was the same as in the experimental preparation. First, a control cycle was generated; then a perturbation ("stimulus"), a change of y, was given at some time (old phase) in the next cycle and the program allowed to complete three more cycles. Old phases, times to the succeeding three cycles (cophases), and periods were calculated. The program could be run without noise. When desired it could also be run with pseudorandom noise, using the computer's random number generator to produce small variable changes of x and y that were added or subtracted to the x and y variables generated by the equation for each of the 7616 times per cycle that they were computed.

The top five panels of Fig. 3 show plots of the noiseless model of phase resetting with different strengths of stimulation. They are consistent with the various phase resetting patterns: Type 1 for weak stimuli (panels I and II, $\Delta y = 1.0$ and 2.1); Type 0 for strong stimuli (panels IV and V, $\Delta y = 3.1$ and 7.0); and a continuous resetting pattern that includes all cophases (panel III, $\Delta y = 2.6063$) for an intermediate strength stimulus given at old phases near 0.5. This represents the phase singularity.

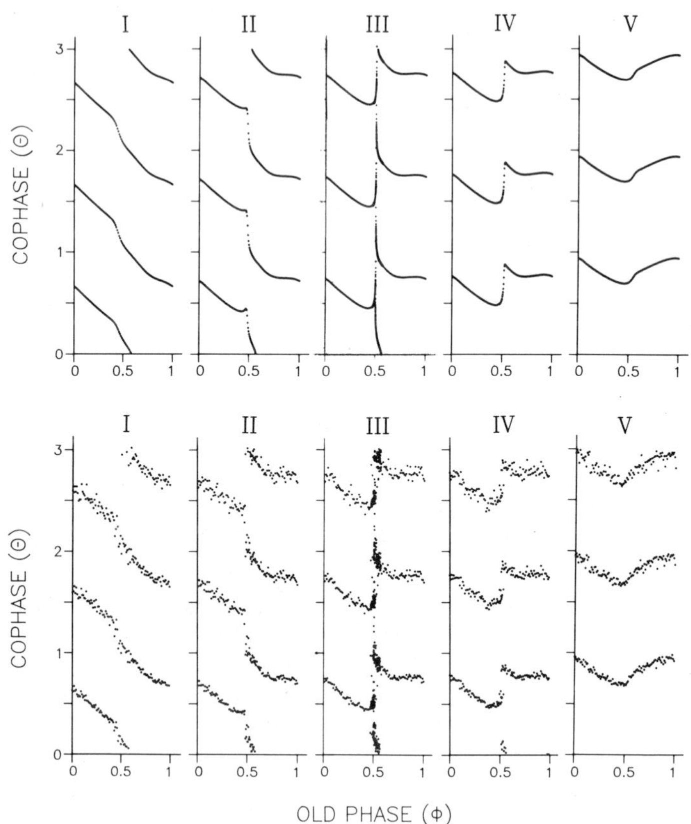

Fig. 3.   Top panels: plots of phase resetting with weak (panel I) to strong (panel V) stimuli in Van der Pol model without noise. Bottom panels: plots of phase resetting with weak (panel I) to strong (panel V) stimuli in Van der Pol model with noise. (Reprinted from Ref. 6)

These results are similar to the experimental findings in the cat (Fig. 1). A three–dimensional plot for the Van der Pol model of cophase as a function of old phase and stimulus strength (depth axis) constructed from the noiseless data in the top panels of Fig. 3 and two additional stimulus strengths ($\Delta y = 0.0$ and 5.0) is shown in Fig. 4. Type 1 resetting appears closest to the observer and Type 0 appears more distant. The plot is a helicoid surface whose boundary follows a corkscrew path around a vertical axis. The axis of the helicoid is intercalated between Types 1 and 0 curves, corresponds to the cophase scatter seen in plot III of Fig. 3, and represents the singularity. The similarity of the model's surface to that of the helicoid constructed from the experimental data obtained in the cat (Fig. 1) is apparent in Fig. 4.

128

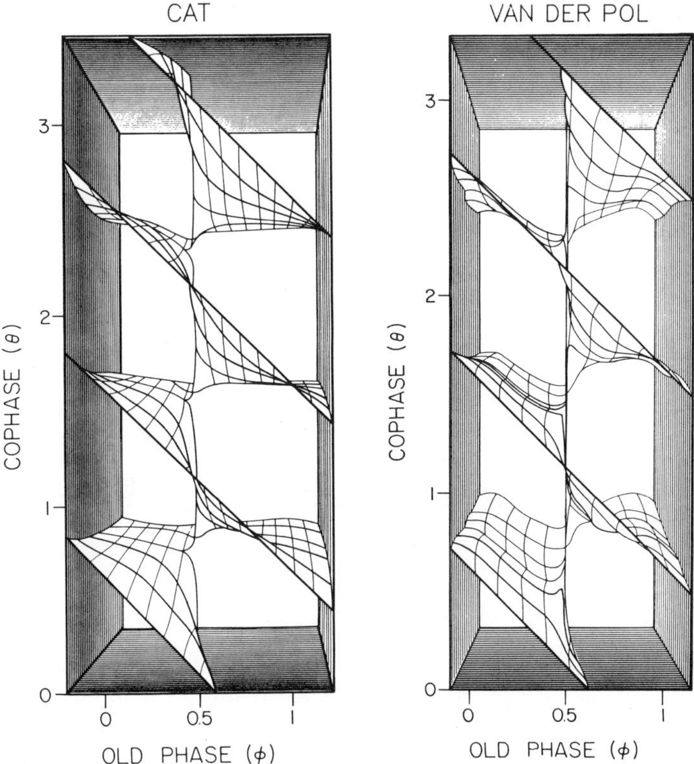

Fig. 4.   Three–dimensional surfaces reconstructed from resetting plots of cat (Fig. 1) and Van der Pol model (Fig.
3). Both surfaces depict helicoid whose boundary follows corkscrew path around vertical axis. (Reprinted
from Refs. 2 and 6)

The addition of a small amount of noise to the model caused minor changes of the stable
limit cycle. However, phase resetting plots were affected by noise as shown in Fig. 3 (lower
panels). All stimulus parameters in these plots are identical to those of the noiseless plots (top
panels). Noise caused slight variations of cohases throughout the cycles with all strengths
of stimuli, and in this aspect caused the model findings to resemble closely the experimental
results (see Fig. 1). Addition of noise during the intermediate strength stimuli (Plot III, lower
panels) changed the resetting pattern from one of continuity, with all cophases represented at
an old phase of about 0.5, to one where at this old phase the cophases were clustered about
certain values. This result also resembles the experimental findings of apparent discontinuity
shown by the example in Fig. 2 (panel III).

A quantitative analysis of this effect in the model shows that while noise causes clustering of cophases in preferred domains, it does not lead to true discontinuity. Figure 5 shows
the findings for the first cycle after a perturbing stimulus. In panel A are plotted the cophases
after 100 stimuli, without noise, at each of six old phases near the singularity ($\Delta y = 2.6063$).
Each different old phase leads to a different cophase, but all 100 trials at each old phase yield
the same cophase. When noise is added, the cophases for the 100 stimuli at each old phase

become dispersed (panel B); although most cophases are represented for each old phase, they clearly cluster in a similar manner for all old phases. This can also be seen (panel C) when percentage distribution of the cophases for the 600 stimuli in panel B are plotted together.

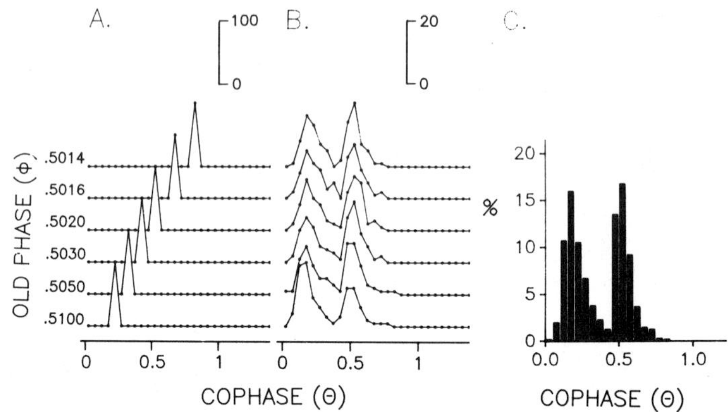

Fig. 5.   Plots of cophase incidence at each of six old phases in the Van der Pol model: A) 100 stimuli at each old phase (no noise); B) 100 stimuli at each old phase (noise); C) distribution of cophases for all six plots of (B). (Reprinted from Ref. 6)

In summary, the findings in the Van der Pol model show that the addition of noise can convert the continuous pattern of phase resetting at the singularity to one in which clusters of resetting times, or cophases, occur in certain time domains. There is, nonetheless, no true discontinuity (Fig. 5) for there are cophases of almost all durations for the first cycle after the stimulus. Because of the clusterings, however, a large number of model stimulus trials was necessary to demonstrate the continuity. Our findings suggest that "noise" may affect the outcome of phase resetting studies on biological preparations. We suggest that in such preparations, where stochastic processes are present and where only a relatively small number of stimulus trials is usually possible, the probability of obtaining all cophases is small. There is thus the appearance, but not the reality, of discontinuities in the resetting plot. The apparent discontinuities, rather than contradicting the limit–cycle hypothesis, are probably explained by the underlying form of the limit cycle, the effects of these stochastic processes and the limited number of points that can be obtained during an experiment[6].

INFLUENCE OF RESPIRATORY "DRIVE"

In the course of our previous experiments in cats, we noted that an increase of respiratory drive; for example, increased respiratory activity produced by hypercapnia could lead to a change of phase resetting patterns even though the parameters of the facilitatory mesencephalic stimulus remained unchanged. We therefore performed additional experimental phase resetting studies in paralyzed cats[7]. In these studies we kept the strength of the mesencephalic stimuli constant but changed respiratory drive by increasing arterial PCO_2, by electrically stimulating a carotid sinus nerve (C.S.N.), or by cooling the intermediate areas of the ventral medulla. Phase resetting patterns were generated before and after each change of drive.

An example of changing drive by hypercapnia is shown in Fig. 6 (left panels). The effect of the increase of drive, with perturbing stimulus strength kept constant, was to convert a

Type 0 (strong) phase resetting pattern at end–tidal PCO_2 of 34 torr to a Type 1 (weak) pattern at a PCO_2 of 50 torr. The right panels of Fig. 6 show that increasing respiratory drive and activity by means of carotid sinus nerve stimulation, while end–tidal PCO_2 remained constant, led to a similar change of resetting pattern from strong to weak.

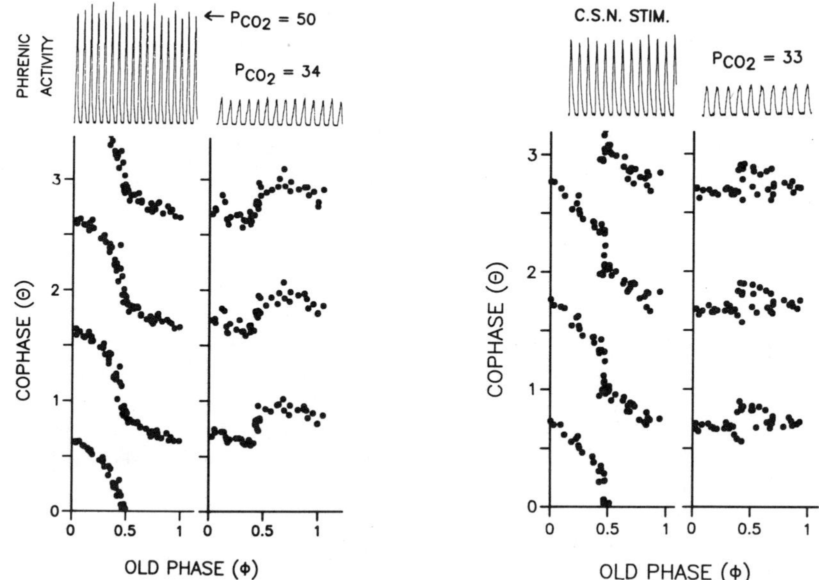

Fig. 6.   Effect on phase resetting patterns of increasing respiratory drive by hypercapnia (left panels) or by stimulation of carotid sinus nerve (C.S.N) at constant PCO2 (right panels). In both, a low level of drive, indicated by the level of phrenic activity, is associated with strong (Type 0) resetting, while a high level of drive is associated with weak (Type 1) resetting (Reprinted from Ref. 7).

Figure 7 shows an example of a cat whose respiratory drive was decreased by localized cooling of intermediate areas of ventral medulla at a constant end–tidal PCO_2. The left panel shows the Type 1 (weak) resetting pattern found when PCO_2 (49 torr) and respiratory drive were high and medullary intermediate area temperatures normal (37.5 °C). When respiratory drive had been reduced by cooling of the intermediate areas to 27 °C, with no change of PCO2, the same perturbing stimulus now acted as a strong stimulus yielding a type 0 resetting pattern (right panel). Raising the local temperature to 29.5 °C led to an increase of drive and phrenic activity (middle panel) and the perturbing stimulus now led to an ambiguous resetting pattern with a wide scatter of cophases, characteristic of singular response at an old phase of 0.5.

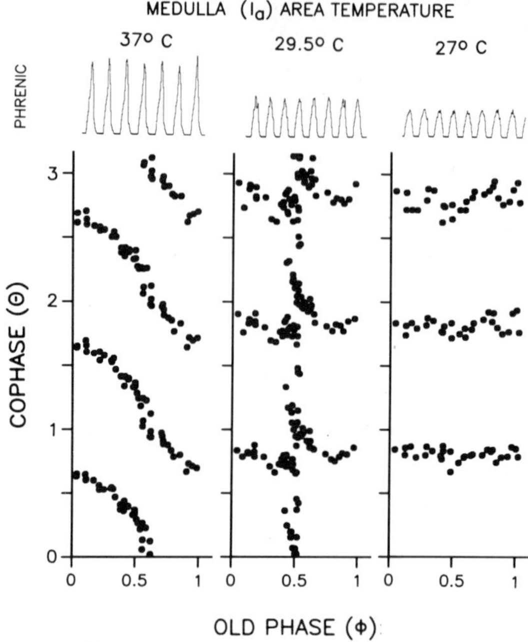

Fig. 7. Effect on phase resetting patterns of changing respiratory drive by cooling intermediate areas (I_a) of medulla. <u>Left:</u> Type 1 (weak) resetting when drive is high. <u>Right:</u> Type 0 (strong) resetting when drive is low. <u>Center:</u> ambiguous re setting pattern, consistent with phase singularity, when medullary temperature and drive are intermediate. $PCO_2 =$ 49 torr in all panels. (Reprinted from Ref. 7)

MODELING OF CHANGES OF RESPIRATORY DRIVE

To model the effect of effects of changing respiratory drives, we modified the Van der Pol equation (see above) by adding a term for "drive" (D) as follows

$$dy/dt = \varepsilon \cdot (1 - x^2/D) \cdot y - x$$

Parameter D defines the size of the Van der Pol oscillator's limit cycle in numerical terms. Increasing D has the effect of increasing the size of the limit cycle in both dimensions without changing its shape (see Fig. 9 below). Phase resetting patterns were generated as described before and were determined for a number of different drive values.

The effects on phase resetting patterns produced by changing the drive parameter of the model are shown in Fig. 8. All five of the resetting plots were generated with the same perturbing stimulus ($\Delta y = 2.6063$) which was chosen to yield an ambiguous pattern at the phase singularity when the drive parameter was 1.0. It can be seen that there is a progressive change in pattern of resetting from Type 0 (right) when the drive was small to Type 1 (left) when the drive level is large, even though the perturbing stimulus strength is the same.

These studies clearly show that a change of respiratory drive, produced by several different types of mechanisms, affects the ability of a perturbing stimulus of constant strength to cause phase resetting of respiratory rhythm; the same stimulus that acts functionally as a strong perturbation at low drive acts functionally as a weak perturbation at high drive. We suggest that the increasing respiratory drive has the effect of increasing the size of the limit cycle of the oscillator; this in turn is responsible for the change of functional effectiveness of the same stimulus and to the change of resetting pattern.

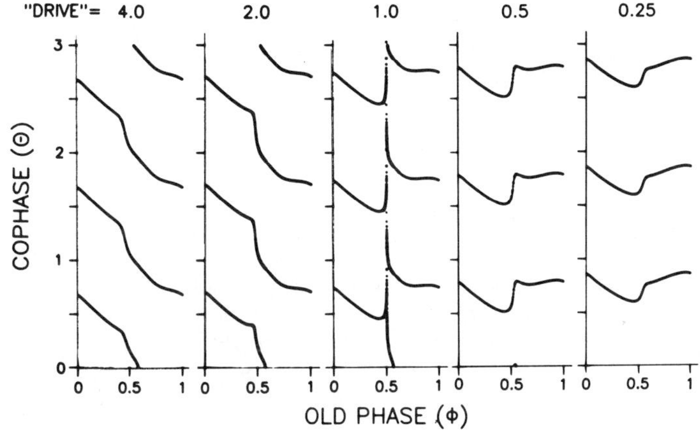

Fig. 8.   Phase resetting plots generated with Van der Pol model with constant perturbing stimulus and different drive parameters. As drive increases (from right to left), resetting patterns change from Type 0 to Type 1 with a singular response at an intermediate drive level of 1.0 (center). (Reprinted from Ref. 7)

The phase resetting patterns with changing drive in the Van der Pol model were qualitatively similar to the experimental, and further support the conclusion, since the stimulus was provably constant in the model. On theoretical grounds, increased drive and limit–cycle size should affect the function strength of perturbations. This is shown in Fig. 9. Top panels depict the limit cycle generated with a small D of 0.25 and the effects of five perturbations of equal strength ($\Delta y = 3.0$) given at various times in the cycle. Because of the small size of the limit cycle, all initial points after the perturbation are shifted so that they do not enclose the phaseless points ($x = 0$, $y = 0$); Type 0 resetting therefore results[4]. Bottom panels show the limit cycle generated with a large D of 4.0. Because the limit cycle is large the initial perturbed points, although shifted the same absolute magnitude ($\Delta y = 3.0$) as with the smaller drive, do enclose the phaseless point; a Type 1 resetting pattern is therefore generated.

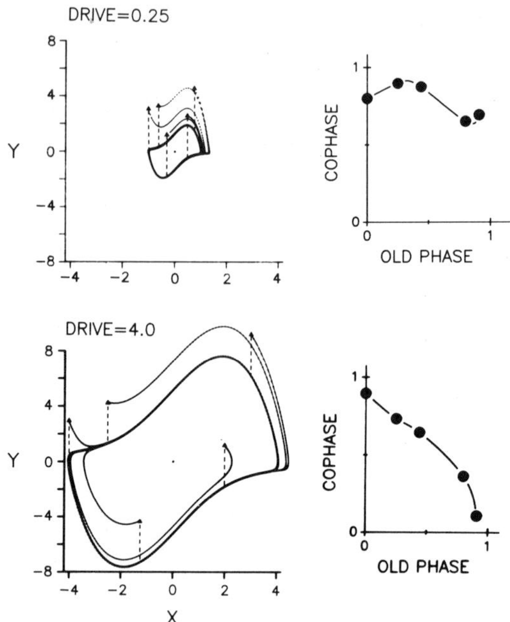

Fig. 9. Top panels. Van der Pol limit cycle generated at low drive with 5 separate perturbations given at various times in cycle (left). Because the perturbations do not enclose phaseless set, a Type 0 (strong) resetting pattern results (right). Bottom panels: limit cycle with high drive and five perturbations (left) of same magnitude as above. Because these perturbations do enclose the phaseless point, a Type 1 (weak) resetting pattern results. (Reprinted from Ref. 7).

CONCLUSIONS

In summary, we have shown that changing drive in both in vivo respiratory experiments in cats and in the Van der Pol oscillator model leads to changes in the functional effectiveness of the same perturbing stimulus due to changes in the size of the limit cycle. The findings should apply to any limit–cycle oscillator, but they do lend further support to the idea that a continuous limit–cycle mechanism underlies the generation of respiratory rhythm in the mammal. It is suggested that in phase–resetting experiments where the effects of a perturbing stimulus are being studied, it is important that drive be kept constant. This is especially true for respiration where drive can easily change.

ACKNOWLEDGEMENT

This work was supported by USPHS Grant HL–17689.

REFERENCES

1. A. T. Winfree, "The Geometry of Biological Time," Springer–Verlag, New York, (1980).

2. D. Paydarfar and F. L. Eldridge, Phase resetting and dysrhythmic responses of the respiratory oscillator, Am. J. Physiol. 252:R55 (1987).

3. D. Paydarfar, F. L. Eldridge and J. P. Kiley, Resetting of mammalian respiratory rhythm: existence of a phase singularity, <u>Am. J. Physiol.</u> <u>250</u>:R721 (1986).

4. L. Glass and A. T. Winfree, Discontinuities in phase–resetting experiments, <u>Am. J. Physiol.</u> <u>246</u>:R251 (1984).

5. B. Van der Pol, On "relaxation–oscillations," <u>Phil. Mag.</u> <u>2</u>:978 (1926).

6. F. L. Eldridge and D. Paydarfar, Phase resetting of respiratory rhythm in a model of a limit–cycle oscillator: Influence of Stochastic Processes, <u>in</u>: "Respiratory Control: a Modeling Perspective", G. Swanson and F. Grodins, eds, Plenum, New York (in press).

7. F. L. Eldridge, D. Paydarfar, P. G. Wagner and R. Thomas Dowell, Phase resetting of respiratory rhythm: effect of changing respiratory "drive," <u>Am. J. Physiol.</u> <u>257</u>: R271 (1989).

DISTINGUISHING RANDOM FROM CHAOTIC BREATHING PATTERN BEHAVIOR

Stanley M. Yamashiro

Biomedical Engineering Dept., University of Southern
California, Los Angeles, CA 90089-1451

Observations of breathing pattern behavior often show seemingly random characteristics during spontaneous breathing[1]. Experiments in animals also indicate that important
non-linearities are involved in the respiratory rhythm generator[2,3,4]. The question that can be
asked is whether this randomness represents noise or chaos produced by a non-linear system[5].
This paper reviews the methods of analyzing such data to answer this question including the
newer topological (geometric) techniques based on fractal analysis.

TIME DOMAIN ANALYSIS

Since breathing can be identified in terms of events (inspirations or expirations), the
time between events can be analyzed in terms of inter-event histograms, mean, and variance.
Comparison of these measures with known random processes such as the Poisson is one way of
studying fluctuations. However, this approach cannot be used to distinguish a random from a
chaotic process. This is because it is possible to duplicate statistical measures corresponding
to a Poisson random process with a chaotic process[6].

FREQUENCY DOMAIN ANALYSIS

Analyzing the power spectra of records of airflow or volume versus time reveals spectral
peaks and a general broadband nature[1]. Again, such characteristics can be shared by random
or chaotic processes. By varying the settings of a mechanical ventilator, it is possible to
entrain the respiratory oscillator at different frequencies[4]. Power spectral density analysis
could then be used to study how the entrainment frequency varies as a function of different
forcings. For a chaotic process, a sequence of bifurcations of solutions is commonly observed
where the period of oscillation doubles as a parameter is changed until an infinite period
chaotic regime is reached[6]. Thus, power spectra could be used to study such potential routes
to chaos. Figure 1 shows the time response of a Van der Pol oscillator to a sinusoidal velocity
forcing. The corresponding power spectral density is shown in Figure 2. A spectral peak
is seen at about 0.1 Hz which is about half the forcing frequency which was set at .25 Hz
(15 bpm). The Van der Pol oscillator has been found by Eldridge[7] to exhibit similar phase
resetting properties as the respiratory oscillator.

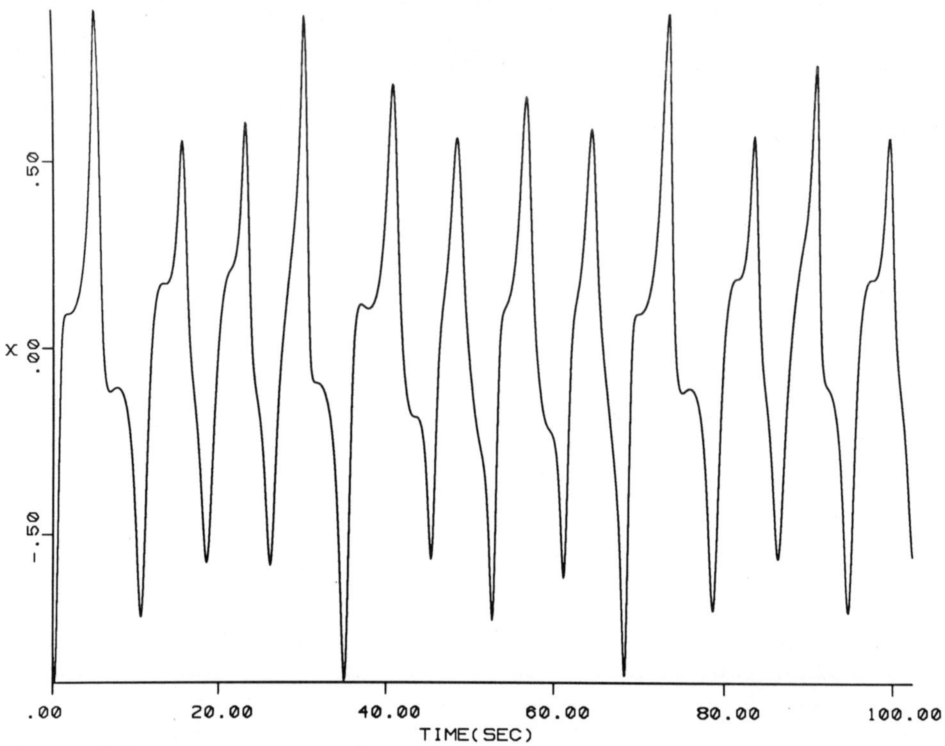

Fig. 1. Time response of a Van der Pol oscillator to velocity forcing.

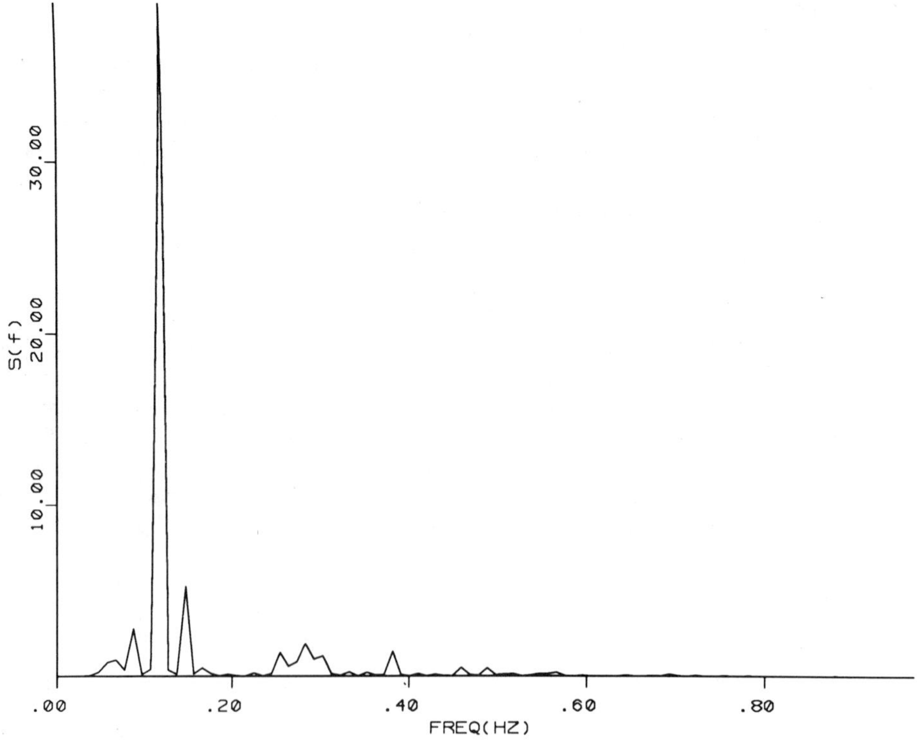

Fig. 2. Power spectral density of a Van der Pol oscillator.

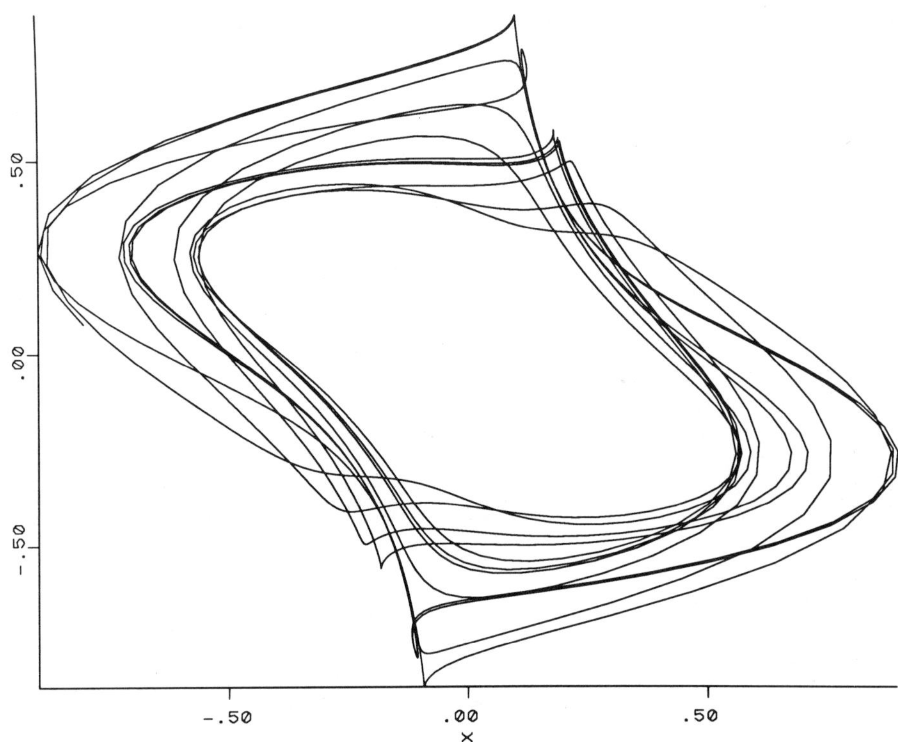

Fig. 3. Phase space portrait of a Van der Pol oscillator.

TOPOLOGICAL ANALYSIS

Topological analysis is based on phase space plots of the independent state variables as suggested by Poincare. Choice of the proper state variables for breathing pattern analysis is uncertain, but reasonable choices would include volume and airflow. Figure 3 shows a phase space plot of velocity versus position for the Van der Pol oscillator corresponding to the same conditions as Figures 1 and 2. Note that while trajectories do not superimpose, they do show a pattern and remain in the vicinity of a nominal trajectory. Even if only a single variable such as volume or airflow is measured, a phase space analysis can be accomplished by sequentially delaying a given time series record by a fixed lag L. An M-dimensional vector is then created as:

$$x(t), x(t + L), \ldots, x(t + (M - 1)L) \tag{1}$$

Figure 4 shows this procedure applied to the Van der Pol oscillator for 2 dimensions and a lag of 24 samples. The trajectories are different than the true phase portrait shown in Figure 3, but it does show a similar property of remaining in the vicinity of a nominal trajectory. By the proper choice of L, independent dynamic variables can be produced which approximate or embed the topological properties of the original process. The mathematical justification for this procedure leads to the requirement that: $M > 2K + 1$ where K equals the true order of the state vector[8]. Thus, a 2 dimensional state vector requires an embedding dimension M greater than 5 to satisfy this inequality.

139

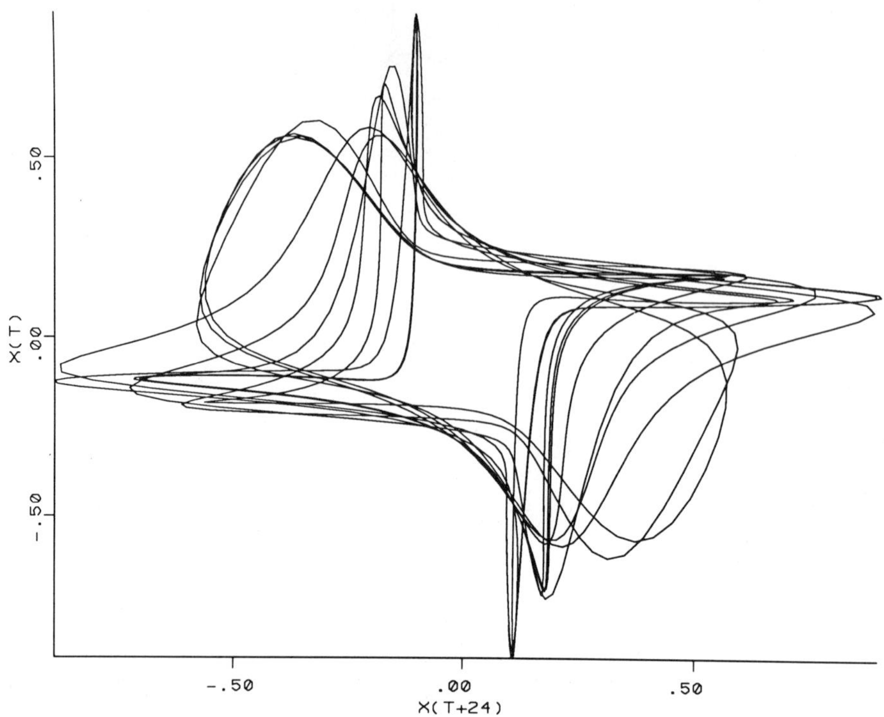

Fig. 4. Phase space portrait using 2 dim. embedding.

The reason for transforming time series data to the phase space lies in the general property of many chaotic processes to have characteristic patterns which emerge from the dynamic trajectories despite having random like behavior. These patterns are best summarized by fractal analysis[9] and this is the best current way of discriminating chaos from noise.

FRACTAL ANALYSIS

Fractal analysis originated out of a desire to describe natural shapes like coastlines and trees which show shape independence of scale, usually referred to as self-similarity. If a coastline is measured using a ruler exactly one meter long, a total length in arbitrary units (e.g. cm) L_m is obtained. If a ruler one km long is used instead, a shorter length L_{km} is measured in the same units due to the inability of measuring some of the irregularities. The two lengths can be related as:

$$L_{km} = L_m S * *(1 - D) \tag{2}$$

where S = scale (meters/km) and D = fractal dimension. Typical coastlines have a D of about 1.25 which would lead to $L_{km}=.25\ L_m$. Fractal dimension can be experimentally estimated from plots of log L versus log S. This approach has been found useful in describing the change in relative dispersion of myocardial blood flow measured by microspheres as a function of tissue sample size[10].

CORRELATION INTEGRAL

An M-dimensional signal consists of M*N points where N=the number of samples used for analysis. The Euclidian distance can be computed between each pair of points as well as the correlation integral:

$$C(v) = (\text{No. pairs}(I, J)\text{whose dist.}(x(I) - x(J)) < v/N_{tot}) \tag{3}$$

where N_{tot} is the total number of pairs considered[5]. Fractal analysis is then based on plotting $\log C(v)$ versus $\log v$ and the slope used to estimate fractal dimension D. Since the embedding dimension M can be arbitrarily chosen, D can be estimated as a function of M. For random noise, D will always equal M since white noise has infinite dynamic dimension. A chaotic process will instead result in D reaching a plateau and remaining fixed as M is increased. This is shown in Figure 5 which shows the estimated D as a function of M for the Van der Pol oscillator previously described. A plateau is reached at about a dimension of 2.6 for an embedding dimension of 7.

Figure 6 shows time series data collected for skeletal muscle vasomotion in terms of microvessel red cell velocity[11]. Vasomotion is a phenomenon with random appearing characteristics which may be caused by relaxation oscillator type behavior of vascular smooth muscle.

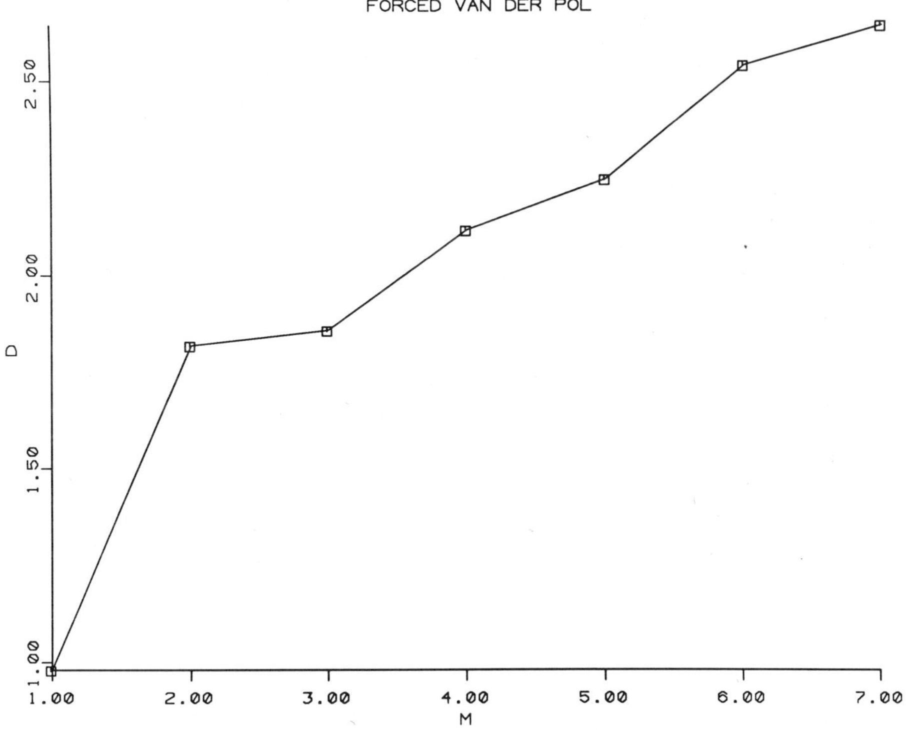

Fig. 5. Fractal dimension vs embedding dimension for a Van der Pol oscillator.

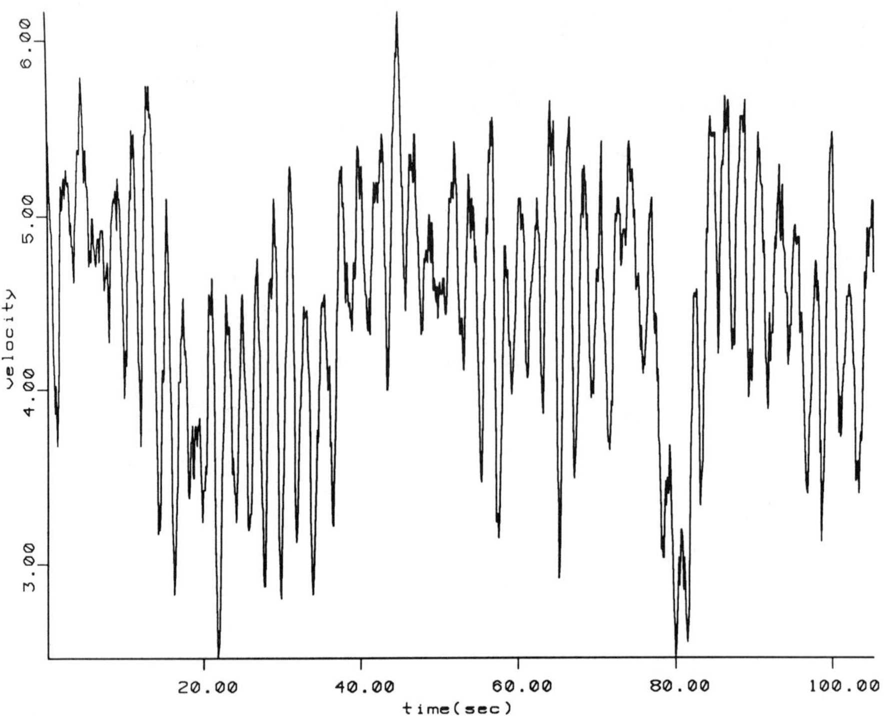

Fig. 6. Skeletal muscle red cell velocity vs. time.

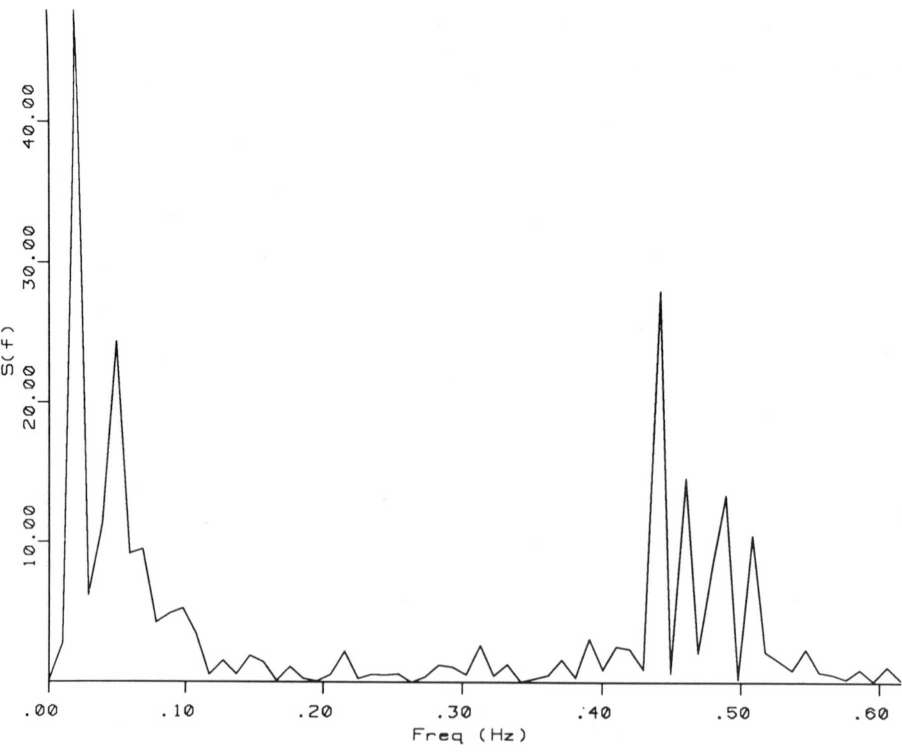

Fig. 7. Power spectum of vasomotion.

The power spectrum of the record in Figure 6 is shown in Figure 7. Two peaks are present, one at about .05 Hz and the other at .45 Hz. Note that the highest spectral peak occurs at a frequency which is at least half of heart rate, similar to the forced Van der Pol oscillator shown in Figure 2. The two dimensional phase plane plot using a delay of 5 samples is shown in Figure 8. Not much structure can be seen from this figure, but a general elliptically shaped trajectory is suggested when observing the time evolution of the points. The estimated fractal dimension for vasomotion data is shown in Figure 9 which indicates a plateau at a dimension of 4.0. Thus, vasomotion does not appear to involve random noise and a chaotic process is suggested by fractal analysis.

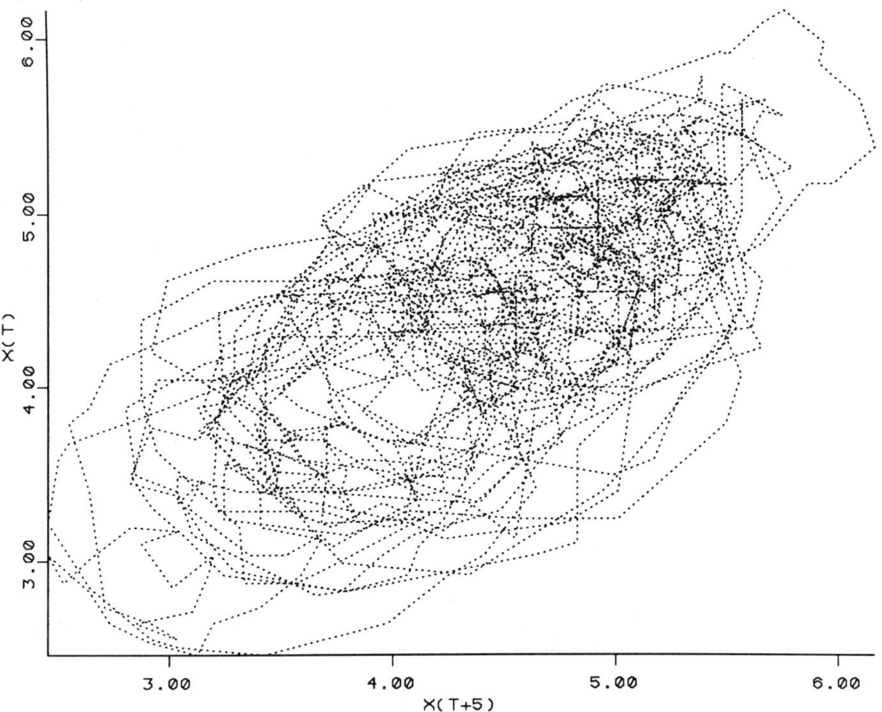

Fig. 8. Phase plane portrait of vasomotion.

CONCLUSIONS

Of the currently available methods of analysis, the most promising for distinguishing random from chaotic behavior are spectral and topological analyses. Estimation of fractal dimension from correlation integral measurements is a logical first step for uncovering chaotic behavior. Breathing pattern and vasomotion appear to involve relaxation oscillators of the Van der Pol type which respond to forcing in a way resembling the bifurcating route to chaos. Thus, manipulating the magnitude and frequency of a periodic forcing should be a useful tool for studying potential chaotic regions.

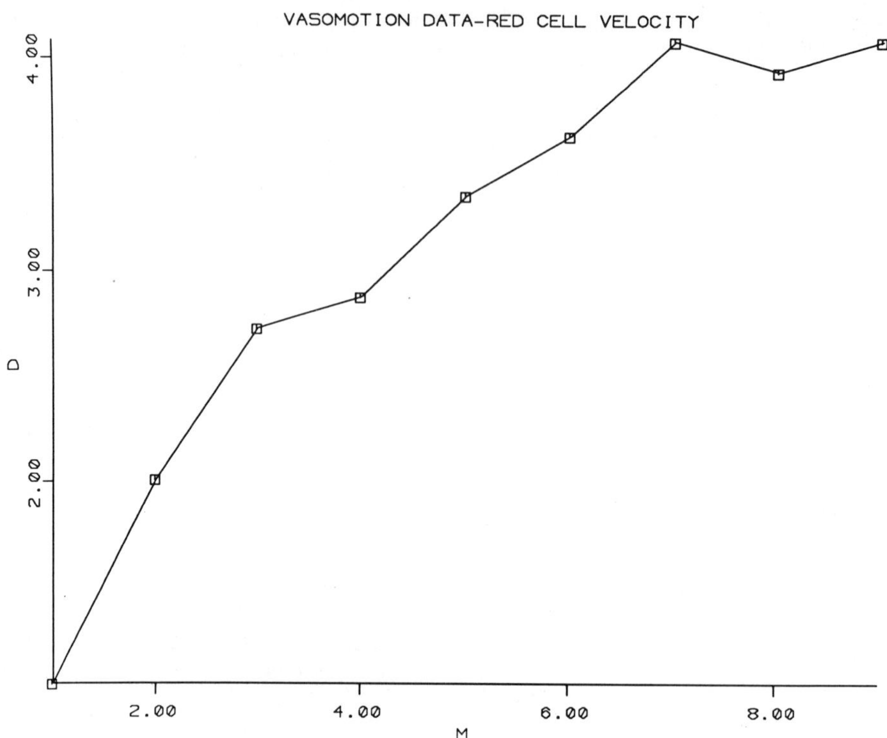

Fig. 9. Fractal dimension vs. embedding dimension for vasomotion.

REFERENCES

1. L. Goodman, Oscillatory behavior of ventilation in resting man, <u>IEEE Trans. Biomed. Eng.</u> <u>BME-11</u>:82-93 (1964).

2. D. Paydarfar, F.L. Eldridge, and J.P. Kiley, Resetting of the mammalian respiratory rhythm: Existence of a phase singularity, <u>Am. J. Physiol.</u> <u>250</u>:R721-R727 (1986).

3. D. Paydarfar and F.L. Eldridge, Phase resetting and dysrhythmic responses of the respiratory oscillator, <u>Am. J. Physiol.</u> <u>252</u>:R55-R62 (1987).

4. G.A. Petrillo and L.A. Glass, Theory for phase locking respiration in cats to a mechanical ventilator, <u>Am. J. Physiol.</u> <u>246</u>:R311-R320 (1984).

5. P. Grassberger and I. Procaccia, Measuring the strangeness of strange attractors, <u>Physica.</u> <u>9D</u>:189-208 (1983).

6. L. Glass and M.C. Mackey, From clocks to chaos, <u>in</u>: "The Rhythms of Life," Princeton Univ. Press, Princeton, (1988).

7. F.L. Eldridge, Phase resetting of the respiratory oscillator-experiments and models, <u>in</u>: (This volume)

8. J.M.T. Thompson and H.B. Stewart, "Nonlinear Dynamics and Chaos," Wiley, New York (1986).

9. B. Mandelbrot, "The Fractal Geometry of Nature," W.H. Freeman, San Francisco (1983).

10. J.B. Bassingthwaighte and J.H.G.M. Van Beek, Lightning and the heart: fractal behavior in cardiac function, Proc. IEEE 76:693-699 (1988).

11. Data provided by D.W. Slaaf and J.B. Bassingthwaighte.

FOREBRAIN MECHANISMS RELATED TO RESPIRATORY PATTERNING

DURING SLEEP-WAKING STATES

Ronald M. Harper

Department of Anatomy and Cell Biology and the Brain Research Institute
University of California, Los Angeles, Calif.

Classical models of control systems for breathing are traditionally based on concepts of oscillatory networks located in medullary or spinal CNS regions, and principally consider influences from metabolic demands, chemical transduction, and thoracic sensors as control parameters. These models are appropriate for the most basic notions of what duties the respiratory control system is supposed to perform, i.e., provide oxygenation and get rid of wastes, but they fail to appreciate a vast array of functions that the respiratory system serves in real life, as opposed to conditions imposed on an anesthetized, paralyzed, temperature-controlled or perhaps decerebrate preparation mounted in a stereotaxic frame or similar restraining device. Moreover, these models provide little assistance in determining the mechanisms causing the dramatic changes in respiratory patterning accompanying different sleep and waking states, changes that are out of all proportion to alterations in metabolic demands during different states.

The functions that respiratory control systems perform derive partially from our heritage; our ancestors used a set of locomotive muscles to simultaneously provide both movement and a system for providing airflow, albeit air dissolved in a fluid, for survival. After migrating to a silicon environment, some structural changes occurred, but an examination of our latter-day anatomy suggests that the original organization has not really changed very much; we have a set of abdominal and diaphragmatic musculature that greatly aids locomotion and posture, particularly when standing, shifting weight, or actively walking, by maintaining thoracic pressure and abdominal wall rigidity in such actions. The use of respiratory musculature to maintain thoracic pressure in motor control assumes particular prominence in some species; a leaping cat, for example, will inspire before its launch, then activate its laryngeal abductors to close the upper airway and provide an air-filled thoracic shock absorber for landing. An infant in respiratory distress with impending fatigue of the diaphragm or abdominal muscles will also retain air in the thorax by using the much-less-energy-demanding laryngeal muscles; the resulting "grunting" respiratory pattern is thus remarkably energy efficient, but the normal oscillatory models for respiratory control do not readily accommodate these actions. Thus, respiratory control systems and locomotor systems are strongly interdependent, and mutually interact for reasons other than simple metabolic demands placed on the respiratory system by motility.

Respiratory and locomotor systems are so tightly interactive that slow-wave electrical activity, recorded from a rostral brain structure and highly predictive of motor acts, also appears in bursts coincident with inspiratory and expiratory action of respiratory musculature. Rhythmical slow-wave activity in the 4-8 Hz band can be recorded from the hippocampus of many small animals. Activity in the lower range of this band (4-6 Hz) is associated with tonic immobility or "automatic" movements such as grooming. Activity in the faster portion of the band (6-8 Hz) is associated with "voluntary" movements[1] (Fig. 1). At least two populations of neuronal systems appear to mediate electrical activity at these two frequency bands, and these systems appear to use two separate neurochemical modulators.[1] The envelope of activity recorded from the hippocampus bears a time dependency to the respiratory cycle, with the phase dependent on respiratory period. Hippocampal rhythmic slow-wave activity

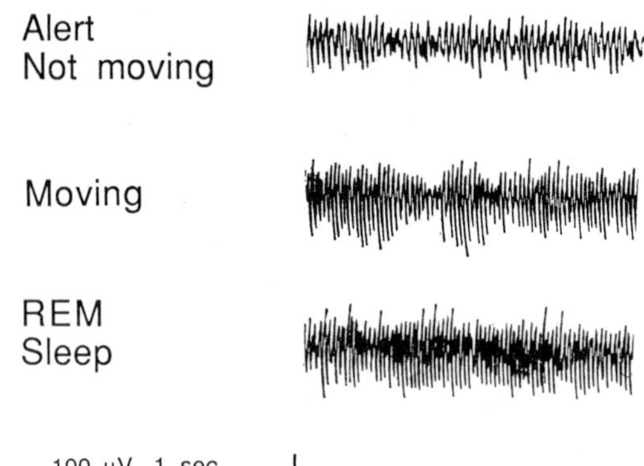

Fig. 1.   Hippocampal rhythmical slow-wave activity during immobility, active movement, and rapid eye movement (REM) sleep in the rabbit. Note the near-sinusoidal rhythmical slow-wave activity and the change in frequency with different behaviors. Amplitude of this activity is modulated by respiration in the cat. (Adapted from Harper[2])

appears nearly continuously during rapid eye movement (REM) sleep and is of higher frequency than during immobility[2]; these periods of higher frequency activity are associated with more rapid respiration as well. Thus, it would appear that a system that is exhibiting profound slow-wave changes to somatic motor effort is also manifesting remarkable changes in electrophysiological patterns to respiratory patterns. The implication of these findings is that control over somatic musculature and respiratory patterning is closely integrated, and conceptual models of respiratory control must consider this integration.

Respiratory control systems are used by many species to assist in temperature regulation. Raising core temperature has the effect of increasing respiratory rate, followed by a transition to panting, with extreme involvement of upper airway musculature. The strategy of elevating respiratory rate to dissipate heat can be extremely effective (Fig. 2). The neural mechanisms underlying this control lie in the anterior hypothalamus, a rostral brain structure, and a structure in which neurons apparently "dissociate" their normal activity of altering rate with altering core temperature during REM sleep.[3] Thus, REM sleep may "dissociate" rostral temperature-control regions from brainstem regions, although this contention is under active investigation.

Even cortical structures modulate respiratory patterning. Stimulation of the orbital frontal gyrus will cause a switch to the inspiratory cycle, an effect that is abolished during REM sleep[4] (Fig. 3). Again REM sleep has "dissociated" rostral forebrain structures from brainstem structures. A second cortical region, the anterior cingulate cortex, also appears to be related to respiratory patterning; a subset of neurons in that structure discharge with either a breath-by-breath or a tonic discharge relationship to the respiratory cycle. These relationships are state dependent,[5] i.e., the discharge correlations are present in some states but missing in others.

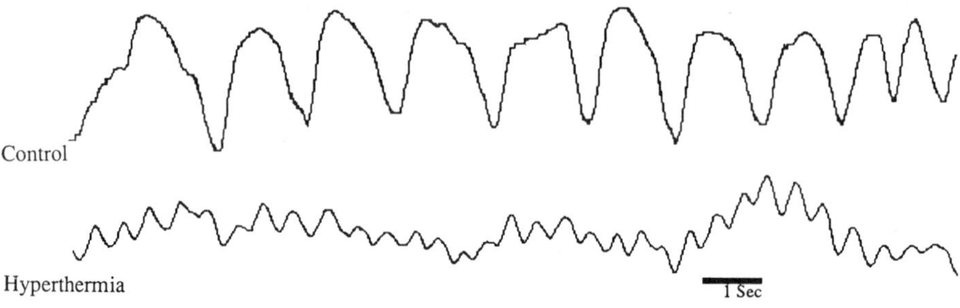

Fig. 2.   Integrated respiratory muscle during waking before and after hyperthermia (induced by cocaine). Note the tachypnea following elevation of temperature.

There exists, in addition to these limbic motor and cortical sites, a limbic arousal system that can provide substantial activation to the upper airway, diaphragmatic, and abdominal motor pools. This system may be particularly active during periods characterized by affect, and is under profound influence of sleep states. The anatomical features of this system have been described by Holstege and his colleagues,[6-8] and some of the functional aspects have been described by Kapp and his co-workers.[9] The limbic structures include, but may not be limited to, the central nucleus of the amygdala and associated nuclear groups such as the bed nucleus of the stria terminalis, and their projections to the so-called pontine "pneumotaxic" area and the nucleus of the solitary tract. Additional projections from the central nucleus of the amygdala terminate in the periaqueductal gray region, which in turn sends fibers to the nucleus of the solitary tract and to laryngeal motor pools (Fig. 4). Evidence for a role in respiratory patterning for all these structures comes from electrical and chemical stimulation, cold blockade, and neuronal recording data. The latter data demonstrate that neurons from these limbic regions discharge on a breath-by-breath basis with the respiratory cycle[10] (Fig. 5), and that this discharge relationship is heavily sleep state dependent, with pattern-dependent discharge occurring in particular sleep states and disappearing in other states. Single-pulse electrical stimulation delivered to portions of this limbic system will entrain the respiratory cycle, a relationship that is also state dependent. Cold blockade of the central nucleus of the amygdala will abolish an aversively conditioned

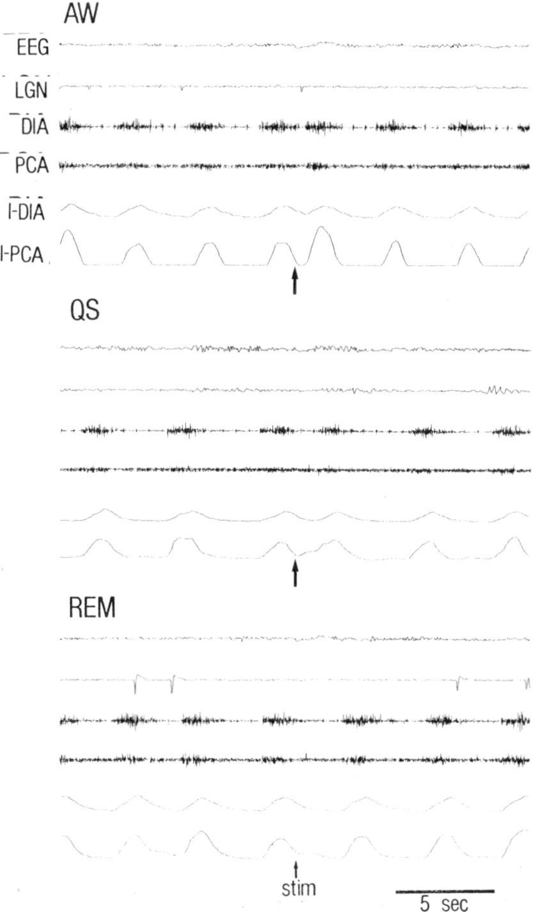

Fig. 3.   Electrical stimulation of the orbital frontal cortex will elicit a switch to inspiration, an effect abolished during REM sleep. EEG, electroencephalogram. LGN, lateral geniculate nucleus electromyogram (EMG). DIA, raw diaphragmatic EMG; PCA, raw posterior cricoarytenoid EMG. I-DIA, integrated diaphragmatic EMG. I-PCA, integrated PCA EMG. (Reprinted with permission from Marks et al.[4])

respiratory response in the waking state.[11] Electrical stimulation of portions of the periaqueductal gray region will elicit vocalization, an arguably important component of respiratory control.[12]

The role that these limbic activation structures play in respiratory patterning is most likely related to functional properties mediating affect, although a more general role is also possible. The state-related influences may be directly related to these affective properties or may be independently influencing the respiratory-related role.

The role that sleep states play in these descending influences is mixed. The potential for "dissociation" of rostral brain influences on caudal brain structures during REM sleep is demonstrable by electrical stimulation and possibly by temperature manipulation, although the latter has yet to be shown. However, some rostral brain structures contain neurons in which the closest breath-by-breath relationship with neuronal discharge is observed in REM sleep, rather than a dissociated effect, and that incongruity has not been solved; perhaps "affective" neurons are particularly activated during REM sleep and show enhanced respiratory relationships. Certainly, sleep states have the potential to modify input from affective contributions. In some states, such as REM sleep, that modification could be substantial. Sleep states may alter patterning of respiratory neurons by modifying transmission properties at the sensory transducer, but this possibility has not been investigated in detail.

In summary, breathing control involves considerably more integration than maintaining an oscillatory pattern in a set of muscles for simple metabolic demands. Substantial contributions from other brain structures impinge on the motoneurons of the upper airway and diaphragmatic musculature, and these contributions include temperature maintenance, motor control, vocalization, and affective demands. Sleep states, in turn, exert control over all of these impinging influences, and those sleep modifiers may not be uniform for each impinging condition; some states may enhance affective contributions, for example, or diminish somatomotor contributions.

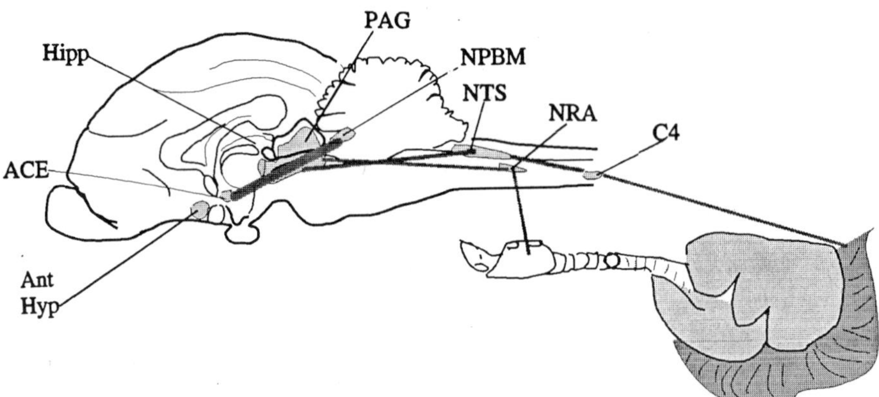

Fig. 4.   Schematic outline of some of the structural features involved in a "limbic arousal" system that influences respiratory patterning. The structures include the bed nucleus of the stria terminalis and the central nucleus of the amygdala (ACE) in the rostral forebrain and their projections to the nucleus parabrachialis medialis (NPBM) of the pons, the periaqueductal gray (PAG), and the nucleus of the solitary tract (NTS). The PAG, in turn, contains neurons that influence nucleus retroambiguus (NRA) cells, premotor to upper airway and abdominal musculature involved in vocalization. In this drawing, NRA is shown innervating the laryngeal musculature. The affective projections via the NPBM and NTS affect the phrenic pool in C4, motoneurons for the diaphragm. The hippocampus (Hipp) may assist in modulating skeletal motor activity, of which the respiratory musculature is a part. The anterior hypothalamus (Ant Hyp) provides descending temperature "drive" to respiratory musculature.

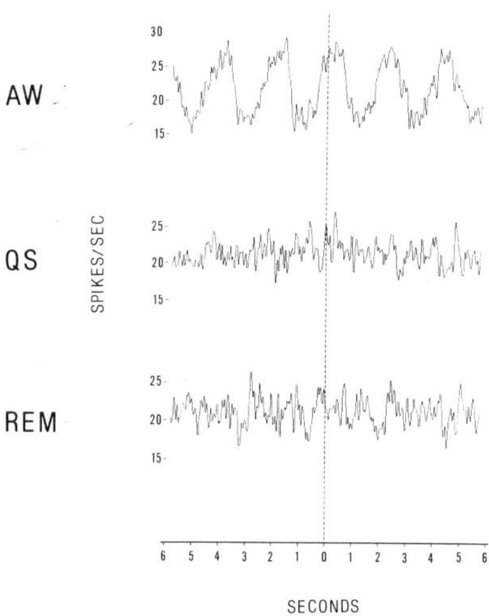

Fig. 5. Cross-correlation histogram of a single neuron in the central nucleus of the amygdala during three sleep-waking states. Note the pronounced correlation during waking (AW) that was attenuated during quiet sleep (QS) and absent during rapid eye movement sleep (REM). (Adapted from Zhang et al.[10])

ACKNOWLEDGMENTS

Supported by grant HL22418 from the National Heart, Lung and Blood Institute. Dr. Robert Trelease and Ms. Karen Kluge provided substantial contributions to this manuscript.

REFERENCES

1. C. H. Vanderwolf, Hippocampal electrical activity and voluntary movement in the rat, Electroencephalogr. Clin. Neurophysiol. 26:407-418 (1969).
2. R. M. Harper, Frequency changes in hippocampal electrical activity during movement and tonic immobility, Physiol Behav. 7:55-58 (1971).
3. P. L. Parmeggiani, A. Azzaroni, D. Cevolani, and G. Ferrari, Responses of anterior hypothalamic-preoptic neurons to direct thermal stimulation during wakefulness and sleep, Brain Res. 269:382-385 (1983).
4. J. D. Marks, R. C. Frysinger, and R. M. Harper, State-dependent respiratory depression elicited by stimulation of the orbital frontal cortex, Exp. Neurol. 95:714-729 (1987).
5. R. C. Frysinger and R. M. Harper, Cardiac and respiratory relationships with neural discharge in the anterior cingulate cortex during sleep-waking states, Exp Neurol. 94:247-263 (1986).
6. D. A. Hopkins and G. Holstege, Amygdaloid projections to the mesencephalon, pons and medulla oblongata in the cat, Exp. Brain Res. 32:529-547 (1978).
7. G. Holstege, L. Meiners, and K. Tan, Projections of the bed nucleus of the stria terminalis in the mesencephalon, pons, and medulla oblongata in the cat, Exp. Brain Res. 58:379-391 (1985).
8. G. Holstege, Some anatomical observations on the projections from the hypothalamus to brainstem and spinal cord: an HRP and autoradiographic tracing study in the cat, J. Comp. Neurol. 260:98-126 (1987).
9. B. S. Kapp, M. Gallagher, R. C. Frysinger, and C. D. Applegate, The amygdala, emotion and cardiovascular conditioning, in: "The Amygdaloid Complex," Y. Ben-Ari, ed., Elsevier, Amsterdam (1981).
10. J. X. Zhang, R. M. Harper, and R. C. Frysinger, Respiratory modulation of neuronal discharge in the central nucleus of the amygdala during sleep and waking states, Exp. Neurol. 91:193-207 (1986).
11. J. X. Zhang, R. M. Harper, and H. Ni, Cryogenic blockade of the central nucleus of the amygdala attenuates aversively conditioned blood pressure and respiratory responses, Brain Res. 386:136-145 (1986).
12. G. Holstege, Anatomical study of the final common pathway for vocalization in the cat, J. Comp. Neurol. 284:242-252 (1989).

Part IV:

SLEEP AND RESPIRATORY CONTROL STABILITY

CHEMORECEPTION IN SLEEP

Jerome A. Dempsey, James B. Skatrud and Kathe G. Henke

Departments of Preventive Medicine and Medicine
University of Wisconsin-Madison, Madison, WI

Wakefulness has a significant influence on ventilatory control. We have examined some aspects of this wakefulness influence by investigating the effects of sleep: a) on CO_2 retention and respiratory muscle function; and b) on periodic breathing.

Alveolar Hypoventilation and Respiratory Muscle Recruitment

During NREM and REM sleep, hypoventilation occurs in all humans. The explanation seems straightforward enough, based on the fundamental work by Orem et al.[1] with unanesthetized, tracheostomized cats, which showed a reduction in phasic medullary neuronal activity coincident with the loss of wakefulness. The problem with this explanation is that in the human with intact upper airways, this hypoventilation is not accompanied by decreased motor output to the respiratory muscles. To the contrary, diaphragmatic, and especially scalenus muscle EMG activity is increased significantly in all stages of NREM sleep relative to wakefulness. With progression into deeper stages of NREM sleep, these muscle activities increase even more and abdominal expiratory muscle EMG becomes phasically active[2,3]. Mechanical correlates of these increased EMG activities include increased trans-diaphragmatic pressure and increased end–expiratory gastric pressure during sleep[3,4]. We reasoned that this apparent paradox of increased respiratory muscle activity coincident with alveolar hypoventilation may have been due to a sleep–induced increase in airway impedance. Indeed, upper airway resistance does increase substantially with sleep, and the increase is progressive with depth of sleep[4].

We tested this hypothesis in normal snoring and non–snoring subjects by reducing airway resistance via either nasal CPAP or by breathing a low density gas (He:O_2). CO_2 was either permitted to change during these reductions in airway resistance, or was controlled by altering inspired CO_2 concentrations. Ventilatory output was significantly increased by reducing resistance 30%–50% during NREM sleep and P_{ETCO_2} was reduced to within 1–2 mmHg of waking levels. With low airway resistance maintained via CPAP and P_{ETCO_2} restored to normal sleeping levels, ventilatory output was increased 60%–70% greater than was obtained in the normal, i.e. high resistance, sleep. These substantial effects of reducing airway resistance were significant– although less in magnitude–in non–snorers as well as in snorers.

Modeling and Parameter Estimation in Respiratory Control
Edited by M.C.K. Khoo
Plenum Press, New York

When resistance was reduced via CPAP or He:O_2 and $PaCO_2$ fell, phasic respiratory muscle EMG activities were decreased or eliminated. Expiratory abdominal and inspiratory scalenus EMG activities were most senstitive to these combined chemical and mechanica' changes, and diaphragmatic EMG was least sensitive. When CO_2 was restored to norm:

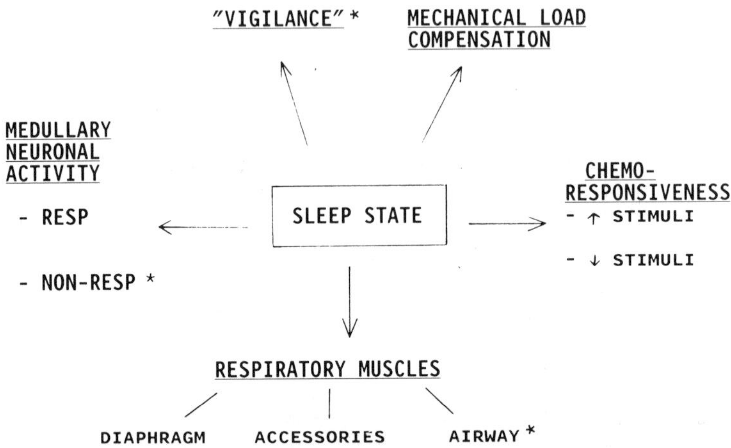

Figure 1. Potential effects of removal of wakefulness on the control of breathing. As discussed in the text, three major effects of sleep are relevant to the question of sleep–induced hypoventilation: 1) reduced phasic and tonic activity of "non-respiratory" medullary neurons, leading to; 2) reduced activation of upper airway abductor muscles and therefore increased airway resistance. The third critical effect of sleep is loss of "vigilance", secondary to cortical inhibition which is primarily responsible for the absence of significant (immediate) compensatory response to the mechanical load of increased airway resistance. Thus, V_T falls and hypoventilation occurs when resistance increases in NREM sleep. Chemoresponsiveness to CO_2 is not changed via sleep state alone, thus phasic respiratory muscle EMG activity increases (in the steady state) during sleep; and this effect is greater on accessory inspiratory and expiratory muscles than on the diaphragm. Upper airway muscle activity is also recruited by increased CO_2 so that resistance is reduced during sleep as CO_2 rises, but still remains substantially greater than awake values. An additional significant effect of the NREM sleep state is the profound inhibition of ventilation and respiratory muscle phasic activity when P_{aCO_2} is reduced a few mmHg, i.e. sleep unmasks an apneic threshold (see below).

sleeping levels while airway resistance was maintained at a low level (via CPAP), EMG activities were increased about halfway (abdominal), most of the way (scalenus), or all of the way (diaphragm) back to eupneic (i.e. high resistance) levels as obtained in NREM sleep. These changes in EMG activity under conditions of reduced airway resistance and $PaCO_2$ restored to normal eupneic (sleeping) levels, indicate the importance of the resistive load, per se, to respiratory muscle recruitment during sleep.

What are the implications of these data for defining the "wakefulness" stimulus to breathing? Combining the findings of Orem et al. in the tracheostomized cat with ours in the intact human, we propose that while there is clearly some depressant effect of sleep on medullary inspiratory neurons, the major effect of sleep is on "non–respiratory" medullary neuronal activity which would reduce phasic and tonic upper airway muscle activity and increase airway resistance. It is this secondary effect of the increased resistance which causes most of the hypoventilation because the compensatory response to mechanical loading in sleep is markedly compromised[5,6]. In turn, the diaphragm, and to a greater extent the accessory inspiratory and expiratory muscles, are recruited in response to the increased CO_2 primarily, and to a smaller but significant extent in response to the mechanical load, per se. Certainly the loss of wakefulness does exert some direct effect on phasic respiratory medullary neurons, but we think this would contribute a relatively small amount to the observed hypoventilation and account for only minor amounts of the sleep–induced reductions in CO_2 and hypoxic responses that are sometimes observed in sleep. We see no reason to believe that chemoresponsiveness, per se, is affected by NREM sleep. The relative effects of a secondary effect of sleep (via increased airway resistance) vs a primary effect (via reduced medullary inspiratory neuronal activity) on ventilatory output would probably differ from subject to subject according to the amount of sleep–induced increase in airway resistance.

Finally, we emphasize that sleep does cause very specific state–dependent effects on respiratory muscle function. Dick et al[7] have shown a state–dependent effect on increasing intercostal expiratory muscle activity which is progressive as sleep stage deepens throughout NREM, and is not dependent upon a changing upper airway resistance. Furthermore, in REM sleep, phasic and tonic EMG activity of inspiratory accessory muscles (scalenus and sternoclidomastoid) is markedly reduced or abolished (at least in COPD patients), which may explain some of the reduced FRC commonly observed in REM sleep[8].

Periodic or Unstable Breathing

An example of periodic breathing during sleep produced in the healthy subject upon hypoxic exposure is shown in Figure 2.

Some time ago we proposed[9], based on our observations and the models of Cherniack and Longobardo[10] that this type of periodic breathing occurs in sleep because of a combination of three factors. First, apnea occurs during NREM sleep when $PaCO_2$ is reduced via active or passive hyperventilation by only 3–5 mmHg, i.e. to waking levels, as would occur with hypoxic stimulation of ventilation during sleep[11,12]. Secondly, chemoreceptor responsiveness is maintained during NREM sleep. Thirdly, ventilatory response to hypoxia follows a curvilinear relationship and is interactive with increasing $PaCO_2$. So, with a hypoxic–induced hyperventilation in sleep, a waxing and waning tidal volume first appears. This is followed by apnea and large augmented breaths following each apneic period (Figure 2). Ventilatory inhibition is attributed to the sensitive hypocapnic–induced apneic threshold; the augmented "cluster" breaths are attributed to the high gain of the peripheral chemoreceptors responding to an asphyxic stimulus. Among healthy persons the amount of periodic breathing in hypoxic sleep is positively correlated to the magnitude of the ventilatory response to asphyxia while awake[22].

Subsequent work has shown that this scheme is probably too simplified and a number of other important mechanisms may contribute to this periodicity:

1. After discharge phenomenon: a stabilizer.

Eldridge et al.[13] electrically stimulated the carotid sinus nerves in anesthetized cats causing phrenic nerve activity to increase. When they suddenly stopped carotid sinus nerve stimulation, phrenic nerve activity returned only very slowly to prestimulus baseline activity. If this after–effect is indeed present following physiologic types of stimuli then this potentiated "after discharge" mechanism would prevent ventilatory inhibition from occurring and no periodic breathing would occur. Accordingly, the apnea of hypoxic–induced periodic breathing might then be attributed to the suppression of this after discharge by long–standing cerebral hypoxia, as recently suggested by Younes[14]. All this intriguing possibility needs is the clear demonstration of a significant after discharge of ventilatory output following a truly physiologic stimulus applied during sleep.

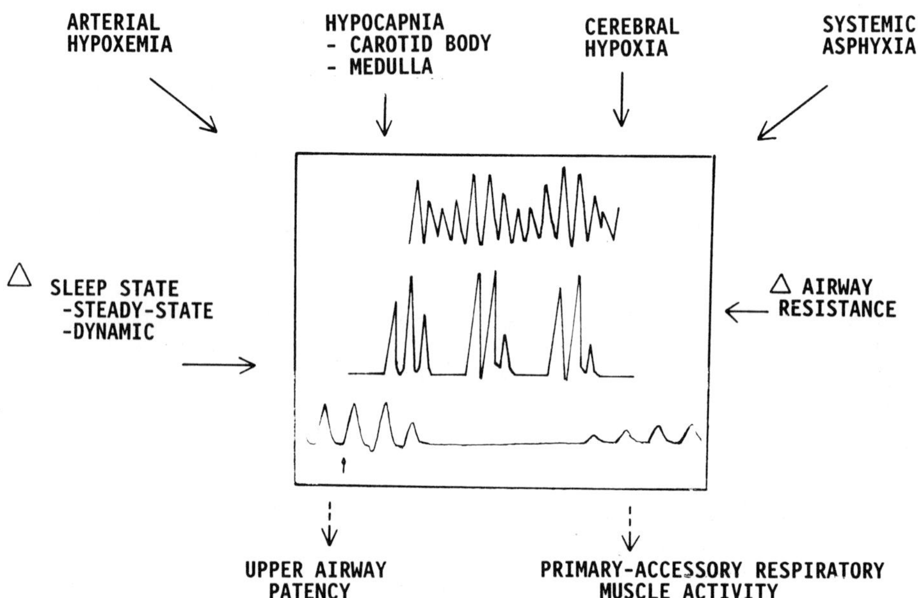

Figure 2. Causes of hypoxic–induced periodic breathing in sleep. The examples shown here occurred during NREM sleep and include, the waxing and waning of V_T early during hypoxic exposure (top), frank periodic breathing with an 20 sec cycle length (middle), and the reduced V_T and prolonged T_E which occurs when hypoxic hyperventilation is suddenly relieved with hyperoxia (bottom). The solid arrows refer to the factors which may cause or change periodic breathing (see text), and the broken arrows refer to some of the consequences of periodic breathing.

2. The importance of tonic inputs to ventilatory phase switching.

To "switch" from inspiration to expiration requires a critical, minimum level of tonic input to the respiratory controller, so that the inspiratory "ramp generator" may be activated at regular intervals. Several factors might reduce tonic input sufficiently so that periodic breathing simply becomes a "switching" problem. These factors might include hypocapnia–acting at peripheral and medullary chemoreceptors–as described above. Cerebral hypoxia

156

may also play a role as Holtby et al.[15] recently showed a depressed ventilation following > 20 minutes of isocapnic hypoxia in awake subjects. We have also recently observed a significant prolongation of T_E following 3–5 mins of isocapnic hypoxia during sleep. Unfortunately, it is difficult to determine whether this apparent effect of cerebral hypoxia is truly independent of cerebral hypocapnia, because the latter occurs secondary to increased cerebral blood flow, even when systemic isocapnia prevails[16]. An apparent contradiction–at least to conventional thinking–is the existence of periodic breathing in hypoxia when carotid chemoreceptor input was reduced via carotid body denervation or by dopamine infusion[17]. These authors speculate that periodic breathing may not be dependent on peripheral chemoreceptors; but is rather an inherent central rhythm exposed by the suppression of "tonic" drives. Reduced metabolic rate, i.e. $\dot{V}_{CO_2}$ or $\dot{V}_{O_2}$, may also cause destabilization of breathing pattern through reduction in tonic drive, as shown by the central and obstructive apneas in the untreated hypothyroid patient during NREM sleep[18]. Perhaps the destabilization effects of hypoxia may also be linked to a reduced $\dot{V}_{O_2}^{17,19}$. Unfortunately, we have no idea as to receptor site or nervous pathways through which a change in metabolic rate may exert its effect on ventilation or breathing pattern. Finally, it is also to be expected that the assumption of the sleeping state will also contribute to a significant reduction in tonic input to the medullary controller[1]. Of course, not all reductions in tonic input cause unstable breathing. We produced highly significant amounts of sustained systemic metabolic alkalosis via sodium bicarbonate infusion over many hours in humans (increase plasma $[HCO_3]$ 6–8 meq/l). This caused significant hypoventilation to the same extent during both wakefulness and sleep, but the stability of breathing pattern was not disturbed.

3. Dynamic changes in sleep state.

A dynamically changing sleep state may be necessary for the perpetuation of periodic breathing by being responsible for the generation of very large breaths at the point that each apneic period was broken. At least a portion of this augmented breath may be because of the increased gain attributable to the rise in P_{CO_2} from below to above the apneic threshold, combined with the synergistic effect of asphyxic stimuli acting on the carotid body (see above). It is also likely that this apparent increase in gain is attributable to a microarousal immediately prior to, or coincident with, these initiations of the breath. This coincidence may be shown by the simultaneous analysis of the time course of changing ventilatory output vs EEG spectral frequency. Quite often the EEG change preceeds the initiation of ventilatory output, especially in very light sleep stages, as shown in some elderly subjects[20]. Much work is needed on this important question, as current analyses suffer from EEG motion artifacts and as yet improperly defined changes in "sleep state". The effect of a true change in state may affect ventilatory response because of a change in controller gain and also because of a reduction in airway resistance[21]. The most profound effect of state change is observed in REM sleep, where periodic breathing in hypoxia does not occur[9]. Perhaps the tonic input to the respiratory controller is so elevated in REM that switching mechanisms are not sufficiently suppressed by hypoxia and/or hypocapnia. The sensitivity to hypocapnia needs testing during REM sleep.

4. Importance of Airway impedance

When elevated in sleep, airway impedance may have a significant influence on periodic breathing by preventing instability. A sufficiently high resistance may actually prevent periodic breathing, possibly because the impedance does not permit sufficient hyperventilation and hypocapnia to occur. In snoring subjects, we were rarely able to produce periodic breathing by administering hypoxia during NREM sleep[21]. This effect may also explain why periodic breathing in hypoxic sleep usually occurs in stages lighter than 3–4, as the airway resistance is usually considerably lower in these lighter sleep stages. Andrews, Johnson, and Symonds[17] also reported a higher incidence of hypoxia–induced periodic breathing in sleep in lambs when airway resistance was reduced via bypassing the upper airway with a reversible tracheostomy.

In summary, we conclude that a number of factors beyond the powerful combination of hypocapnic inhibition and high carotid chemoreceptor gain will potentially influence periodic breathing in sleep. Each of these potential mechanisms deserves further study. Nevertheless, in the intact, sleeping human, we believe these oscillating stimuli at the very least, provide an essential amplification of any underlying primary oscillation which may exist.

ACKNOWLEDGEMENTS

We wish to thank Jennifer Thomas for manuscript preparation. This work was supported by NHLBI and The Council for Tobacco Research.

REFERENCES

1. J. Orem, I. Osorio, E. Brooks and T. Dick, Activity of respiratory neurons during NREM sleep, J. Neurophysiol. 54:1144–1156 (1985).

2. J.A. Dempsey, K.G. Henke and J.B. Skatrud, Regulation of ventilation and respiratory muscle function during sleep in: "Proceedings in the 1st International Symposium on Sleep and Respiration," F. Issa, P. Surratt, and J. Remmers, eds., Banff, Alberta, Canada (in press).

3. J.B. Skatrud, J.A. Dempsey, S. Badr and R. L. Begle, Effects of high airway impedance on CO_2 retention and respiratory muscle activity during NREM sleep, J. Appl. Physiol. 65:1676–1685 (1988).

4. J.B. Skatrud and J.A. Dempsey, Airway resistance and respiratory muscle function in snorers during NREM sleep, J. Appl. Physiol. 59:328–335 (1985).

5. S. Badr, R. Begle, J. Skatrud and J. Dempsey, Effect of mechanical loading on expiratory and inspiratory muscles during NREM sleep, Am. Rev. Resp. Dis. 137:127 (1988).

6. C. Iber, A. Berssenbrugge, J.B. Skatrud and J.A. Dempsey, Ventilatory adaptations to resistive loading during wakefulness and non–REM sleep, J. Appl. Physiol. 52:607–614 (1982).

7. T.E. Dick, P.L. Parmeggiani and J. Orem, Intercostal muscle activity during sleep in the cat: an augmentation of expiratory activity, Resp. Physiol. 50:255–265 (1982).

8. M.N. Johnson and J.E. Remmers, Accessory muscle activity during sleep in chronic obstructive pulmonary disease, J. Appl. Physiol. 57:1011–1017 (1984).

9. A. Berssenbrugge, J.A. Dempsey and C. Iber et al., Mechanisms of hypoxia–induced periodic breathing during sleep in humans, J. Physiol. (Lond) 343:507–524 (1983).

10. N.S. Cherniack and G.S. Longobardo, Cheyne–Stokes breathing: an instability in physiologic control, N. Engl. J. Med. 288:952–957 (1973).

11. J. Skatrud and J. Dempsey, Interaction of sleep state and chemical stimuli in sustaining rhythmic ventilation, J. Appl. Physiol. 55:813–822 (1983).

12. K.G. Henke, A. Arias, J.B. Skatrud and J.A. Dempsey, Inhibition of inspiratory muscle activity during sleep, Am. Rev. Resp. Dis. 138:8–15 (1988).

13. F.L. Eldridge, Central neural respiratory stimulating effect of active respiration, J. Appl. Physiol. 37:723–735 (1974).

14. M. Younes, The physiological basis of central apnea and periodic breathing, <u>Current Pulmonology</u> <u>10</u>:256–326 (1989).

15. S. Holtby, D. Bereganski and N.R. Anthonisen, The effect of 100% O_2 on hypoxic eucapnic ventilation, <u>J. Appl. Physiol.</u> <u>65</u>:1157–1162 (1988).

16. A. Berkenbosch and J. DeGoede, Effects of brain hypoxia on ventilation, <u>Eur. Resp. J.</u> <u>1</u>:184–190 (1988).

17. D.C. Andrews, P. Johnson and M. Symmonds, Metabolic rate and periodic breathing in the developing lamb, <u>Proceedings of the Physiological Society–Oxford Meeting</u>, Oxford, England, 141P (1989).

18. J. Skatrud, C. Iber, R. Ewart, G. Thomas, H. Rasmussen and B. Schultze, Disordered breathing during sleep in hypothyroidism, <u>Am. Rev. Resp. Dis.</u> <u>124</u>:325–329 (1981).

19. H. Gautier, M. Bonora and J. Remmers, Effects of hypoxia on metabolic rate of conscious cats during cold exposure, <u>J. Appl. Physiol.</u> <u>67</u>:32–38 (1989).

20. G. Warner, J. Skatrud and J. Dempsey, Effects of hypoxic–induced periodic breathing during sleep on upper airway obstruction, <u>J. Appl. Physiol.</u> <u>62</u>:2201–2211 (1987).

21. A.I. Pack, M.F. Cola, A. Goldszmidt and D.A. Silage DA, Coherent oscillations in ventilation and frequency content of the electroencephalogram, <u>J. Appl. Physiol.</u> (in press).

22. K.R. Chapman, E.N. Bruce, B. Gothe, and N.S. Cherniack, Possible mechanisms of periodic breathing during sleep, <u>J. Appl. Physiol.</u> <u>64</u>:1000–1008 (1988).

POTENTIAL CAUSES OF RECURRENT APNEAS DURING SLEEP

Neil S. Cherniack and Guy Longobardo

Department of Medicine, University Hospitals of Cleveland
and School of Medicine, Case Western Reserve University
2074 Abington Road, Cleveland, OH 44106

INTRODUCTION

The body contains chemoreceptors which respond to changes in PCO_2 and PO_2 and which operate within the respiratory system to keep arterial levels of PCO_2 and PO_2 within fairly narrow limits[1]. Chemoreceptors send signals to neurons in the brain which, in turn, actuate respiratory muscles which ventilate the lungs and change the content of O_2 and CO_2 in the blood. Blood is pumped by the heart to the tissues where it is used to metabolize nutrients and provide energy. The blood is then returned to the lungs where the process repeats itself. Increases in ventilation raise levels of arterial O_2 content and lower CO_2 while the opposite occurs when ventilation is decreased. This system is subject to a variety of disturbances, because of changes in the O_2 and CO_2 content of the inspired air or because of altered levels of metabolism. Other disturbances occur because the respiratory muscles are used not only to breathe but to maintain posture, for temperature regulation, and in digestion. In addition, because respiration can be altered voluntarily, disturbance in blood gas tensions can occur because of this behavioral control when there are changes in light intensity or noise levels in the environment, and when the respiratory muscles are used in speaking or to express emotion.

Usually disturbance produces smooth transitions in ventilation which return blood gas levels of PCO_2 and PO_2 to within normal limits. However, at times the system, when disturbed, produces rather than smooth cyclic changes in ventilation with increases and decreases in tidal volume, frequency and blood gas tensions.

These crescendos decrease and fluctuations in breathing, frequently occurring with interspersed periods of apnea were recognized almost two hundred years ago by Cheyne and Stokes in patients with heart disease and are called Cheyne-Stokes or periodic breathing. Similar changes in breathing have been observed in patients with disease of the brain[2]. The respiratory changes of Cheyne-Stokes breathing are frequently associated with fluctuations in other physiological systems. The changes that occur in association with respiration include waxing and waning of blood pressure, cerebral blood flow, and pupillary size. In these patients who were usually stuporous or comatose, alertness seemed to increase during the phase of hyperventilation while stupor was more profound during apnea. Studies in the past twenty years show that periodic breathing can also occur in apparently healthy individuals exposed to altitude and even under sea level conditions during sleep[3]. Renewed interest in periodic breathing has resulted from the recognition of the sleep apnea syndrome. In this syndrome periodic breathing occurs sometimes with profound decrease in arterial O_2 saturation. One school of thought believes that these profound dips in saturation are themselves dangerous precipitating cardiac arrhythmias or producing over years pulmonary hypertension and failure of the heart. The apneas that occur in the sleep apnea syndrome are of two types: central, in which all respiratory activity ceases or more commonly are obstructive caused by closure of the pharyngeal or laryngeal airway so that air fails to enter the lungs even though respiratory activity continues.

Because it is the apneas that cause desaturation, it is understandable that attention became focused on the apneas rather than on the breathing pattern that flanks the apneic episodes. Often even patients with obstructive apnea display a pattern that strongly resembles periodic or Cheyne-Stokes breathing and it is now recognized that apneas during sleep can have both a central and an obstructive component. The similarities in overall pattern, i.e. periods of increased breathing and apnea occurring both in Cheyne-Stokes breathing and in the Sleep Apnea Syndrome, suggest that at times both varieties of periodic breathing arise through related or common mechanisms. Because the hyperventilation occurring in the sleep apnea syndrome is often associated with arousal just like the hyperventilation of Cheyne-Stokes breathing, it is possible that both types of periodic breathing arise from primarily cyclic changes in alertness.

Another possibility that will be explained in this paper using a mathematical model is that periodic breathing arises from unstable behavior of the respiratory control system. Theoretically, unstable behavior can arise in such systems with sufficiently large disturbances and more easily with certain specific changes in the parameters of the control system such as circulation time (time required for blood to be transported from the lung to the brain), controller gain, and set points. As in linear control systems, increases in information transfer in the system (circulation time) and greater controller gains tend to produce cyclic changes in ventilation rather than smooth transitions after a disturbance. Because the respiratory control system is highly nonlinear, it is not always possible to predict the possible effects of disturbances or changes in the parameters of the system without the use of a rather complex model. The model described in this paper was formulated over twenty years ago but has been modified in the interim based on a variety of experimental observations[4]. It has many similar features to mathematical models described by Khoo et al.[5] and by Grodins[6].

THE MODEL

The mathematical model used in this study comprises a controller, peripheral and central chemoreceptors, and the body, consisting of the circulation and the CO_2 and O_2 stored in the body tissues in physical solution and chemical combination.

The controller attempts to maintain CO_2 and O_2 levels at normal values in the face of disturbances by adjusting ventilation.

Three tissue compartments in which CO_2 and O_2 are stored are included in the model. They are: the lung and its associated arterial blood, or the arterial store; and two tissue stores, the brain and an "all other" compartment, mostly muscle[7].

These compartments are connected in parallel by the circulation, and each has its own tissue volume, blood flow, and metabolic rate.

The controller of the system consists of a central receptor which responds to the carbon dioxide tension of the brain tissue, and a peripheral receptor which responds not only to arterial carbon dioxide tension but also to arterial oxygen tension. Blood flow to the brain is proportional to arterial carbon dioxide tension flowing into the brain. The carbon dioxide and oxygen tensions at the controller sites are assumed to be the average over the previous breath, delayed by the circulation[8].

The system equations follow. Symbols are defined in Table 1.

Table 1. Explanation of Symbols

Gas Tension (mm Hg)

Pa   Instantaneous arterial tension, PaC for CO_2, PaO for O_2

Paex,av   Average arterial tension, over the previous expiration, PaCex,av for CO_2, PaOex,av for O_2

PaCav,d   Average CO_2 arterial tension at the peripheral receptor, averaged over the previous breath, and delayed by the circulation time

PaOav,d   Average O_2 arterial tension at the peripheral receptor, averaged over the previous breath, and delayed by the circulation time

Pv   Instantaneous mixed venous tension, PvC for CO_2, PvO for O_2

Pi   Instantaneous tissue tension for the i-th compartment

PbCav,d   Average brain tissue CO_2 tension at the central receptor, delayed by the circulation time

PB   Barometric pressure

Blood flows (L/min)

$\dot{Q}c$ Cardiac output

$\dot{Q}i$ Blood flow to the i-th tissue

$\dot{Q}b$ Blood flow to the brain

$\dot{Q}m$ Blood flow to muscle

Volume (L)

Va   Arterial tissue volume, including blood and lung tissue

Vi   Volume of the i-th tissue

Vr   Functional residual capacity of the lung

Gas flows (L/min)

$\dot{V}$ Instantaneous ventilation rate, a sinusoid

$\dot{V}$Mag   Amplitude of $\dot{V}$

$\dot{V}$A   Average ventilation rate over a breath

FI   Inspired concentration, FIO_2 for oxygen, $FICO_2$ for CO_2

Metabolic production (L/min)

MRi   Metabolic production of the i-th tissue

Dissociation slopes (L/L/mmHg)

Sb   Dissociation slope for blood, CO_2 or O_2

Si   Dissociation slope of the i-th tissue

The Arterial Store

During Inspiration

$$\frac{d}{dt}\left[\frac{(Vr)Pa}{PB-47} + VaSbPa\right] = \dot{Q}cSb\,(Pv-Pa) + V\left[SW_1\left(\frac{Paex,av}{PB-47}\right) + SW_2FI\right] \qquad (1)$$

where a dead space is assumed, and Paex,av is the average alveolar gas tension from the previous expiration. SW_1 and SW_2 are binary switches and have no physiological significance. $SW_1 = 1$ and $SW_2 = 0$ while gas from the previous expiration is being inhaled from the dead space; after that period, ambient air is inspired and $SW_1 = 0$ and $SW_2 = 1$. FI is the inspired fraction of CO_2 or O_2.

The instantaneous ventilation over a breath is sinusoidal according to

$$\dot{V} = \dot{V}Mag\,\sin 2\pi ft \qquad (2)$$

where $\dot{V}$Mag is the amplitude of $\dot{V}$, and where f, the breathing frequency, can be calculated as

$$f = 0.4\,\dot{V}A + 8.4 \qquad (3)$$

VA is the average ventilation over a breath. $\dot{V}$Mag is calculated as $\dot{V}A/\pi$ since the controller output represents the average ventilation rate over a breath.

During Expiration

$$\frac{d}{dt} \left[\frac{(Vr)Pa}{PB-47} + VaSbPa \right] = \dot{Q}cSb\ (Pv-Pa) - \frac{\dot{V}Pa}{PB-47} \tag{4}$$

The Venous Store

$$\frac{d}{dt}\ (ViSiPi) = QiSb\ (Pa-Pi) + MRi \tag{5}$$

where i indicates either the brain or the "all other" compartment which includes muscle. The dissociation slope for CO_2 of the "all other" compartment is taken as approximating that for water.

Blood flow to brain varies linearly with PaC according to

$$\dot{Q}b = 0.04\ PaC - 0.93 \tag{6}$$

The Controller

The controller output representing ventilatory demand is

$$\dot{V}A = 0.3G_1\ (PaCav,d - G_2) + 0.7G_2\ (PbCav,d - G_3)$$

$$+ \left(\frac{45}{PaOav,d-32.44}\right)\ (PaCav,d-G_2) - G_4 \tag{7}$$

Carbon dioxide and O_2 values at controller sites are assumed to be the average arterial and brain CO_2 tensions of the previous breath, delayed by circulation times between the lung and the chemoreceptors. The first and third terms in Eq. 7 are the contributions of the peripheral receptors, and the term $45/(PaOav,d - 32.44)$ is their oxygen contribution to CO_2 sensitivity. The second term in Eq. 7 is the contribution of the central brain receptor. G_1 is a constant relating ventilation to CO_2 arterial tension. The controller "gain" to CO_2 is represented by the sum of the CO_2 contribution (G_1), and the oxygen contribution ($45/(PaOav,d - 32.44)$). G_2, G_3, and G_4 are constants defining the thresholds of the controller characteristics. Thirty percent of the carbon dioxide drive is considered to originate in the peripheral receptor; 70% in the central receptor. Note that as oxygen tension is reduced the controller characteristic shifts to lower CO_2 tensions, which represents a lowering of the CO_2 threshold for stimulation of breathing. Also, the slope, or gain, of the controller is increased.

If the system is disturbed from its equilibrium point (e.g. by hyperventilation) to a level of CO_2 below the threshold, then a period of apnea results during which CO_2 tension rises and O_2 tension falls. Without the threshold shift and increased gain occurring with lower O_2 levels, breathing would begin at very low rates. However, because PO_2 falls during the apneic period, breathing begins at lower thresholds and higher rates. Thus if the disturbance to the respiratory system produces apnea, the length of the apnea, and hence the severity of the disturbance, will depend on controller threshold. Likewise, controller responses to the same disturbance (the ventilation produced at any given instant in time) will depend both on controller slope and threshold.

The same equilibrium operating point can occur with different gains and different thresholds of the controller. Consequently, the same disturbance, e.g. a period of hyperventilation which reduces PCO_2 to a given subnormal value, will result in different post-hyperventilation apneic periods, and so will differ in severity depending on whether the gain is low or high. To better differentiate the effects of variations of threshold, controller gain, and disturbance on ventilatory response with changes in controller gain, thresholds were calculated using the normal equilibrium controller and multiplying controller output by the increase in gain, thus uncoupling the effects of gain and threshold.

164

RESULTS

When the model is disturbed by a period of hyperventilation, cyclic breathing with interspersed intervals of apnea results similar to that seen in humans with periodic breathing. As shown in Figure 1A breathing gradually stabilizes. Controller gain in the simulation was normal, i.e. 3.0 L/min/mmHg, as was circulation time (6 seconds); resting PCO_2 was 40 mmHg. Less severe hyperventilation which reduced arterial PCO_2 by less hyperventilation (only 15 mmHg) caused cyclic ventilation to occur for a shorter time and without periods of apnea (Fig. 1B). The number of apneic cycles increases with the disturbance severity given the same (normal) controller characteristic.

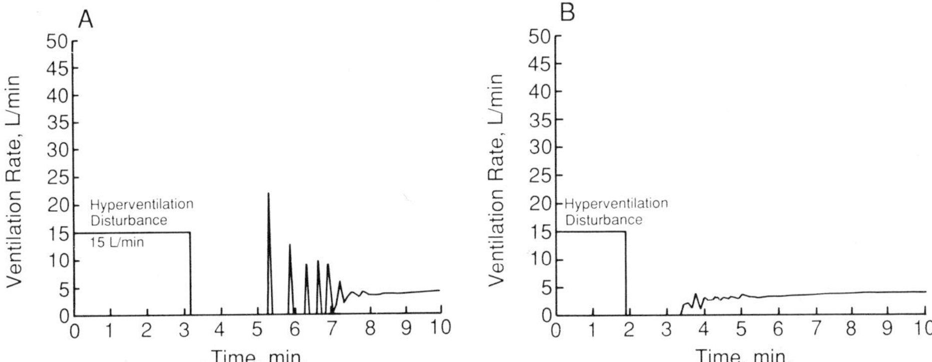

Fig. 1.　Effect of hyperventilation disturbance on apneas. Hyperventilation can reduce arterial PCO_2 enough to cause apnea. When the disturbance is larger (A), cyclic changes in ventilation level occur with apneas. With smaller disturbances (B) the post-hyperventilation apnea period is followed by a rather smooth return of ventilation to its usual level.

The relationship of controller gain to the number of cycles with apnea at several disturbance levels shows, in general, that the greater the controller gain, the greater the number of apneic cycles even when the disturbance is constant. However, the effect of gain is not uniform. For each level of disturbance there is a range of gain changes which can occur without affecting the time during which breathing is periodic.

Circulation time is another important parameter determining the response to a hyperventilation disturbance. With the same disturbance and controller characteristics as in the first simulation, apneic cycles continued for longer times before disappearing as circulation time increased. Eventually cycles became persistent and the system could be considered as perpetually operating in an unstable zone. With greater disturbances, smaller increases in controller gain and less prolonged circulation times are sufficient to cause unstable breathing. The number of apneic cycles also increases after hyperventilation when hypoxic gas mixtures are inhaled.

Periodic breathing occurs most commonly when total metabolic rate is reduced (the brain and the rest of the body) and $PaCO_2$ is elevated as they are during sleep[3]. Instabilities occur more easily at a given controller gain as equilibrium PCO_2 is raised and metabolic rate is reduced, as occurs during sleep. Lowering metabolic rate increases the number of apneic cycles at physiologically achievable controller gains.

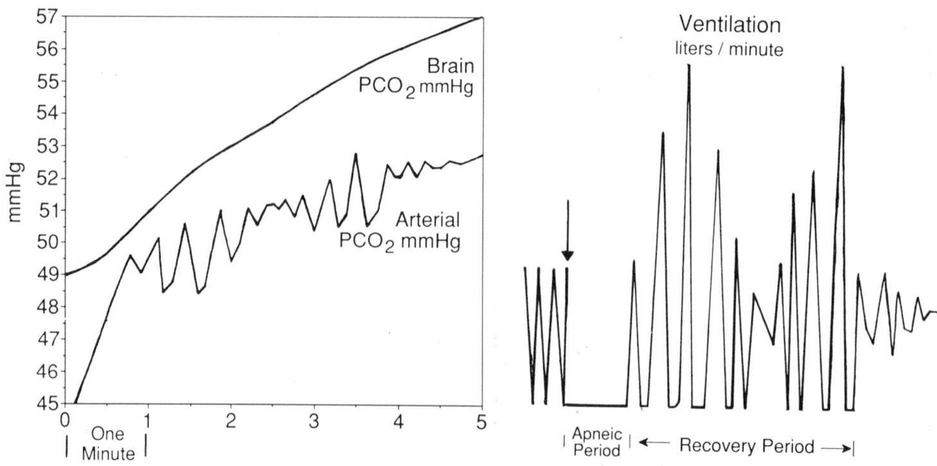

Fig. 2. Time course of changes in ventilation, brain and arterial PCO2 following a shift in controller set point. Note that apnea follows the shift and there is a sustained rise in brain and mean arterial PCO2 levels even when ventilation resumes. Periods of apnea ultimately disappear when arterial and brain PCO2 reach a sufficiently great level. Lowering of brain and/or body metabolic rate increases the time needed for arterial and brain PCO2 to reach the critical level that eliminates apnea.

We next systematically varied only brain metabolism and blood flow to isolate the impact of central chemoreceptors on the stability of respiration (effects of dead space ventilation were not included in these simulations). In the simulations the metabolic rate of the body, excluding the brain, was kept constant. The disturbance was an increase in controller threshold sufficient to shift the operating point of the system so that arterial carbon dioxide tension is increased by 10 mmHg, e.g. from 40 mmHg 'normal' to 50 mmHg. This disturbance results in an apneic period followed by a recovery period during which apneas might occur. Figure 2 shows a simulation of the time course of carbon dioxide tension and ventilation after a change in operating point.

As shown in Fig. 2, ventilation, arterial carbon dioxide and brain carbon dioxide tensions increase after raising the controller threshold. During the apneic period following the disturbance, arterial and brain carbon dioxide tension rise, and arterial oxygen tension falls. When the controller threshold for ventilation is reached, hyperventilation results because of the hypoxia. Since the arterial store is relatively small, arterial gas tensions are easily changed by variations in ventilation. Thus the carbon dioxide tension falls rapidly during the phase of hyperventilation and rises rapidly during the apneic period. When the CO2 tension decreases enough so that the controller threshold for ventilation is reached, another apneic period results. Once again, oxygen tension falls and carbon dioxide tension rises and the cycle continues. But the ability of hypoxia to cause hyperventilation and apnea is limited in both intensity and duration and does not prevent the steady, albeit slow, rise in AVERAGE arterial carbon dioxide tension, an irreversible component about which instantaneous arterial carbon dioxide tension cycles. During this time the brain PCO2 and the larger brain store is steadily rising towards its new equilibrium level, little influenced by the rapid transient CO2 swings in arterial carbon dioxide tension.

Periodic breathing stops when the brain carbon dioxide tension and the irreversible component of arterial carbon dioxide tension combine to reach a level so that hypoxia cannot once again drive blood gases below the apneic threshold. Then breathing is continuous with no apneic periods.

The number of apneic cycles and, concomitantly, the time that apneas persist after a disturbance are not only a function of the magnitude of the brain carbon dioxide tension but also how rapidly it rises to meet the ventilatory threshold level. The rate of change of brain carbon dioxide tension varies directly with cerebral blood flow and inversely with brain metabolism. Therefore, reducing the brain metabolic rate has a multiplicative effect: not only is the ventilatory stimulus arising from the central receptor diminished, but the rise of brain carbon dioxide is slowed, prolonging the time that periodic breathing persists.

Lowering brain metabolism made the number of apneic cycles occurring after a disturbance greater. The number of apneic cycles increased with lowered brain metabolism from three cycles at brain metabolism 1-1/2 times normal to nine cycles when the brain metabolism was 1/2 normal (Table 2). The length of time apneas that persisted during the recovery period also was greater when brain metabolic rate was reduced. The average rate of rise of arterial CO_2 tensions was about the same in each case but brain carbon dioxide tension increased faster when brain metabolism was greater. The central receptor contribution to ventilation was correspondingly greater at any instant of time during the recovery period when brain metabolism was higher. These simulations suggest that changes in brain PCO_2 have a major impact on control system stability. In these simulations cerebral blood flow was held constant at its normal value.

Usually cerebral blood flow increases with hypercapnia[1]. Allowing cerebral blood flow to vary with arterial carbon dioxide tension had a stabilizing effect in model simulations. The number of apneic cycles increased when cerebral blood flow was held constant compared to when it varied with arterial PCO_2. The length of time that apneas persisted also increased when cerebral blood flow was held constant. The results shown in Table 3 are similar for brain metabolic fractions of 0.5 and 1 times normal. Because of the increase in cerebral blood flow, brain PCO_2 becomes closer to arterial PCO_2 with hypercapnia. Raising the set point of the controller hence has a smaller effect on brain PCO_2. Thus the severity of the disturbance to the respiratory system is not as great as when brain blood flow is constant.

The simulation results presented thus far show the influence of reduced brain metabolism in promoting apneas after disturbances. But lowering metabolism of the rest of the body (excluding the brain) can also decrease the stability of the ventilation response. A series of simulations in which brain metabolism was normal and the metabolic rate of the rest of the body varied from 1/3 to 1-2/3 times normal showed increasing numbers of apneic cycles with reduced body metabolism after a threshold disturbance to the controller that raised carbon dioxide operating levels by 10 mmHg (Table 4). The table shows the rate of rise of both arterial and brain PCO_2 increases with this five-fold variation in metabolic rate. Both brain and arterial PCO_2 increase more slowly when body metabolic rates are decreased and this is associated with more apneic cycles.

DISCUSSION

The mathematical model described successfully reproduces periodic breathing. It demonstrates that increases in controller set point or reduced metabolic rates of the brain and body which occur during sleep can precipitate either transient or sustained periodic breathing.

The model also illustrates the importance of controller gains, circulation time, hypoxia, and especially intensity of disturbance in causing recurrent apneas. For example, periodic breathing is known to occur mainly during the deeper stages of NREM sleep in which environmental changes are less likely to disturb ventilation patterns. In addition it suggests that when a disturbance to breathing is prolonged, for example, by an episode of airway obstruction, recurrent apneas are more likely to occur.

Table 2. Number of Apneic Cycles and Time Apneas Persist with Variations in Brain Metabolism. Constant Cerebral Blood Flow .62 liters per minute.

Brain Metabolic Rate Fraction of Normal	Number of Apneic Cycles	Time Apneas Persist minutes	Rates of Rise CO_2 Tension mmHg/min Brain	Arterial
0.5	9	3.0	1.75	.97
1.0	5	1.8	2.05	1.06
1.5	3	0.9	2.40	1.11

Table 3. Number of Apneic Cycles, Time Apneas Persist with Variation in Brain Metabolism and Cerebral Blood Flow. Constant vs. CO_2-dependent Blood Flow

Cerebral Blood Flow lit./min.	Brain Metabolic Fraction 0.5		Brain Metabolic Fraction 1.0	
	Number Apneic Cycles	Time Apneas Persist, min.	Number Apneic Cycles	Time Apneas Persist, min.
.62 Constant	9	3.0	5	1.8
CO_2-dependent Cerebral Blood Flow	3	0.9	0	0

Table 4. Variation in the Number of Apneic Cycles with Changes in Body Metabolic Rate (Excluding the Brain)

	Metabolic Rate Fraction vs. Normal 0.33	Metabolic Rate Fraction vs. Normal 1.67
Number of Apneic Cycles	15	4
Time Apneas Persist (minutes)	4.6	1.4
Rate of Rise CO_2 Tension, Brain	1.1	2.5
Arterial	0.6	1.5

In order to simulate the obstructive apneas, the model would have to include the upper airway muscles. These muscles, like the muscles of the thorax and abdomen which ventilate the lungs, display a respiratory rhythm which responds to changes in CO_2 and O_2. It is believed that during inspiration the thoracic muscles develop negative pressures which collapse the upper airways, but that these occluding forces can be overcome by the forces generated as a result of upper airway muscle activity. We have previously described a model which includes the action of chemoreceptors on upper airway muscles that simulated the events occurring in obstructive apneas[8,9]. It is apparent that any discrepancy in either the amplitude or timing of the upper airway muscle response to CO_2 or hypoxia could precipitate obstructive apneas. Moreover, central apneas by producing blood gas disturbances could trigger recurrent obstructive apneas and vice versa.

We have not included arousal effects in the model. It is clear from experimental observations that the degree of alertness or arousal is affected by changes in PCO_2 and PO_2 though the pathways involved are uncertain. It is possible that changes in blood gas tension triggered by a respiratory disturbance could also affect arousal levels which might intensify or prolong periodic breathing.

Finally, we need to emphasize factors also left out of the model which could affect the occurrence of periodic breathing. These include the relationship between neural activity and ventilation, the effect of memory in respiratory control, possible optimization of respiratory behavior even during sleep to minimize energy costs of breathing, and the possibility of feed forward regulation of breathing (a direct effect of metabolism on the controller).

Thus the model described can only be considered as an initial step but, even at this relatively early stage, indicates that mathematical modeling is a fruitful approach in understanding alterations in respiratory patterns.

REFERENCES

1. N.S. Cherniack and G.S. Longobardo, The chemical control of respiration, Ann. Biomed. Eng. 9:395 (1981).

2. N.S. Cherniack and G.S. Longobardo, Cheyne-Stokes breathing: An instability in physiological control, New Engl. J. Med. 288:952 (1973).

3. E.A. Phillipson, Respiratory adaptations in sleep, Ann. Rev. Physiol. 40:133 (1978).

4. G.S. Longobardo, N.S. Cherniack, and A.P. Fishman, Cheyne-Stokes breathing produced by a mathematical model of the respiratory system, J. Appl. Physiol. 21:1839 (1966).

5. M.C.R. Khoo, R.E. Kronauer, K.P. Strohl et al., Factors producing periodic breathing in humans: A general model, J. Appl. Physiol. 54:644 (1982).

6. F.S. Grodin, "Control Theory and Biological Systems," Columbia University Press, New York (1963).

7. N.S. Cherniack and G.S. Longobardo, CO_2 and O_2 gas stores of the body, Physiol. Rev. 50:196 (1970).

8. G.S. Longobardo, B. Gothe, M.D. Goldman, and N.S. Cherniack, Sleep apnea considered as a control system instability, Respir. Physiol. 50:311 (1982).

9. N.S. Cherniack, G.S. Longobardo, B. Gothe, and D. Weiner, Interactive effects of central and obstructive apnea, Adv. Physiol. Sci. 10:553 (1980).

MINIMAL MODELING OF HUMAN RESPIRATORY STABILITY

David W. Carley

University of Illinois College of Medicine
Chicago, IL 60680

INTRODUCTION

It is known that periodic breathing (PB), or Cheyne–Stokes respiration, may result from a wide range of pathophysiological, metabolic, or environmental conditions including; congestive heart failure, lesions of the central nervous system, hypoxemia, metabolic alkalosis and prematurity[1,2]. Klein first suggested that PB may derive from unstable operation of the chemostatic respiratory control system[3]. It was not until digital computers became widely available, however, that a series of mathematical models was developed to explore Klein's hypothesis[4-11]. Early mathematical models of PB were highly–parameterized and non–linear simulation models[4-8] which served to support the hypothesis that unstable chemical control of respiration, associated with loop gain magnitudes greater than unity, may be common to the wide range of conditions associated with PB. The large numbers of parameters and the difficulties of their laboratory estimation, however, made these models impractical for correlating respiratory control loop gain with observed breathing patterns in individual patients or subjects. More recently, attempts have been made to linearize[9-11] and minimize the parameterization[10-11] of mathematical models of PB.

Carley and Shannon[10] described the first mathematical model of PB in which parameter minimization was explicitly attempted. Their model contains 5 physiologically distinct parameters which may be readily and non–invasively estimated: 1) the ventilatory sensitivity to changes in arterial PCO_2 $(PaCO_2)$, 2) the mixed venous PCO_2 $(PvCO_2)$, 3) the mean lung–volume for CO_2 distribution, 4) the circulatory CO_2 transport capacity and 5) the lung–to–chemoreceptor circulation delay. In the linearized formulation of this model a closed–form expression is developed for $LG(fc)$; the loop gain magnitude at the cross–over frequency. In addition, the relationship between $LG(fc)$ and an impulse–response–based measure of relative stability has been derived[10,12].

It is hoped that this and other minimal models of respiratory stability will help to elucidate the etiology of exaggerated or absent respiratory periodicity under conditions such as sleep–onset, prematurity, generalized autonomic dysfunction, central congenital hypoventilation syndrome, and sleep apnea syndrome. This manuscript describes our initial validation studies of this model under a variety of physiological and pathophysiological conditions.

Modeling and Parameter Estimation in Respiratory Control
Edited by M.C.K. Khoo
Plenum Press, New York

MODEL DESCRIPTION

Figure 1 is a schematic representation of the minimal mathematical model. This is a single–loop proportional control system in which $PaCO_2$ is the feedback, or controlled variable and alveolar minute ventilation ($\dot{V}_A$) is the controller output. The controlled system represents the effects of CO_2 stores and flows in the body, and determines the transfer function between $\dot{V}_A$ as the input and $PaCO_2$ as the output:

$$|B(f)| = \frac{B}{((2\pi f \tau_L)^2 + 1)^{0.5}} \qquad < B(f) = -(\pi + tan^{-1}(2\pi f \tau_L))$$

$$B = \frac{\dot{Q}_C Ks1(Pb - Pw)PvCO_2}{(\dot{V}_A^o + \dot{Q}_C Ks1(Pb - Pw))^2} \qquad \tau_L = \frac{mlv}{\dot{V}_A^o + \dot{Q}_C Ks1(Pb - Pw)}$$

where

$\dot{V}_A^o$:   mean alveolar ventilation
$Ks1$:   slope of CO_2 dissociation curve
mlv: mean lung distribution volume for CO_2
         including gas, tissue, and blood storage
Pb: Barometric pressure
Pw: Water vapor pressure
$\dot{Q}_C$:   cardiac output

This linearized transfer expression assumes that for frequencies relevant to local stability analysis (the critical frequency, which is approximately equivalent to the frequency of periodic breathing), fluctuations in mixed venous PCO_2 do not occur. Thus, the dynamics of tissue gas stores outside the lung need not be described and $PvCO_2$ may be assumed as a constant parameter. It also assumes that $\dot{Q}_C$ and mlv are constant parameters rather than variables. Under these assumptions, a single pole transfer function is obtained where τ_L is the effective time constant of the controlled system.

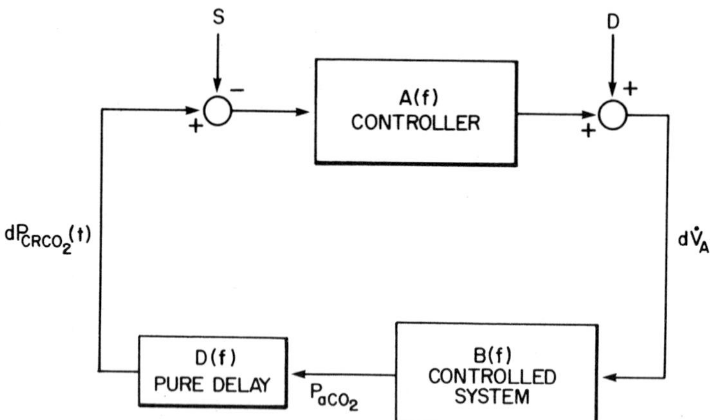

Figure 1. Block diagram of the single loop linear control system model. The controlled system represents the lung–based stores and flows of CO_2, the delay represents circulatory convection from pulmonary capillaries to chemoreceptors, and the controller represents the ventilatory response to changes in $PaCO_2$.

The controller is described by the vector addition of single pole elements representing the effects of central and peripheral chemoreceptors, respectively. The controller impulse response is described by:

$$A(t) = (Ac/\tau c)exp(-(t - D)/\tau c) + (Ap/\tau p)exp(-(t - D)/\tau p)$$

τc: central chemoreceptor time constant
τp: peripheral chemoreceptor time constant
Ac: central chemoreceptor sensitivity
Ap: peripheral chemoreceptor sensitivity
D: circulation delay from lung to chemoreceptor

Notice that the circulation delay (D) is integrated into the above impulse response expression. We will denote by $Ac(f)$ and $Ap(f)$ the frequency–domain representations of the impulse response contributions of the central and peripheral chemoreceptors, respectively. The overall effective CO_2 sensitivity, $A(f)$ is then given by the vector addition of $Ac(f)$ and $Ap(f)$, and the net expression for CO_2 loop gain (LG) is given by:

$$|LG(f)| = |A(f)|\,|B(f)| \qquad\qquad < LG(f) =< A(f)+ < B(f)+ < D(f)$$

where

$$< D(f) = -2\pi f D \quad \text{radians.}$$

A number of the approximations made to achieve this degree of model minimization are readily apparent. Effects of hypoxia, core temperature, and behavioral state on ventilation are not explicitly described, limiting the model to circumstances in which these variables may be approximated as quasi–steady state. Cardiac output, mlv, dead space, and $PvCO_2$ are likewise assumed to remain fixed. Equal circulation delays are assumed between the lungs and both central and peripheral chemoreceptors. A detailed discussion of these and other model assumptions and limitations has been presented elsewhere[10]. Despite these simplifications, however, this model can quantitatively account for many features of the relative stability of respiration in healthy waking and sleeping adults, sleeping patients with familial dysautonomia, and waking patients with congenital central hypoventilation syndrome as described below. The objective of these studies was to evaluate the ability of the minimal model to predict the relative and absolute stability of respiration in individual subjects under a range of environmental and pathological conditions expected to significantly affect the loop gain for CO_2 regulation ($LG(fc)$). This approach was taken as a form of dimensional model analysis to obtain wide ranges in most model parameters, because experimental manipulations to systematically perturb each parameter over a wide range while holding all others constant were infeasible. The basis for comparing model behavior with experimental observations was an index of respiratory oscillations (R) based on power spectrum analysis of ventilation waveforms[12]. Under each condition in each subject, the model R (predicted) was compared to the actual R (observed).

EXPERIMENTAL METHODS

Studies during wakefulness

We have studied 15 healthy awake men (mean age 24 $\pm$ 2 (SD) yr) during progressive hypoxia[12]. Each subject was instrumented for respiration (Respitrace°, calibrated by the method of least squares[13]), CO_2 (Cavitron PM–20NR infrared analyzer with sampling cannula taped 1 cm within the antrum of one naris), transcutaneous oxygen (Radiometer

TCM2), and lead I ECG. Initial measurements included CO_2 excretion rate (5 minute exhaled gas collection), Bohr dead–space, $PvCO_2$ and mlv (rebreathing method of Dubois[14]), and $\dot{Q}_C \cdot Ks1$ (calculated from $PvCO_2$ and $PACO_2$ using the Fick equation). Each of these parameters was then assumed to remain constant for the duration of the study.

During the main protocol, each subject was seated in a comfortable chair while their FIO_2 and $FICO_2$ were adjusted by changing the composition of the 2 l/s bias flow passed through an open–ended hood placed over the head and neck. At each of three FIO_2s, .21, .15, and .12 the values of τc, τp, Ac, Ap, and D were estimated using 3 minute step increases in $FICO_2^{12}$. This allowed estimation of $LG(fc)$ using measured values of all model parameters. In addition, a series of 2 to 3 breath challenges ($FICO_2 = 5\%$) was used at each FIO_2 to perturb ventilation. Similar perturbations were applied to the model (constrained by the estimated parameter values) allowing a comparison of model and physiological behavior. Figure 2 illustrates ventilation and airway PCO_2 for such an "impulse" CO_2 challenge, where $FICO_2$ is increased at the arrow. The ensuing oscillations in both end–tidal PCO_2 and ventilation are evident. The fact that these oscillations decay indicates that the control system is operating in an "under–damped" but stable mode. Figure 3 indicates the expected response types and typical $LG(fc)$ values expected to generate over–damped, under–damped,

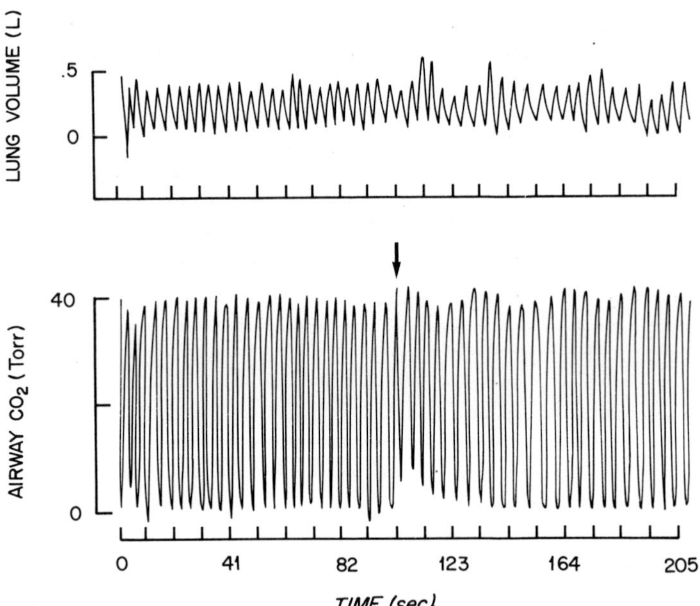

Figure 2. Ventilation and expired PCO_2 waveforms typical of impulse CO_2 challenges used to evaluate relative stability, R. Note oscillations in both expired PCO_2 and ventilation following brief inspiration of CO_2 (arrow).

and unstable operation. Epochs of sustained periodic breathing (PB) were identified by retrospective visual assessment of the polygraphic records and were also characterized by computing R.

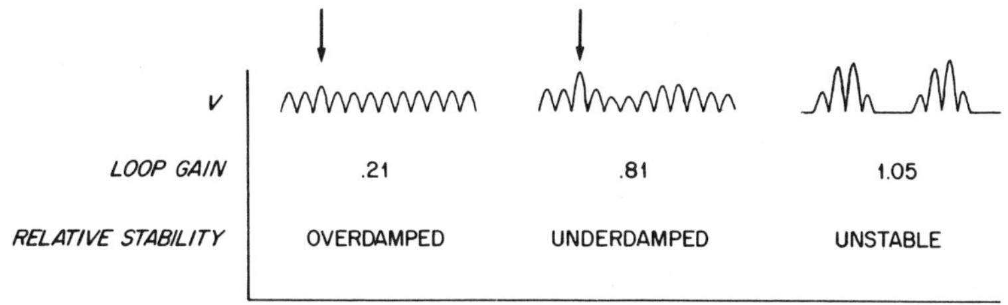

Figure 3. Three distinct modes of control system operation and typical values of $LG(fc)$ for which they would be expected.

Studies during sleep

The transitions from seated to supine posture and from wakefulness to sleep may be expected to induce systematic changes in $PvCO_2$, mlv, D, $\dot{Q}cKs1$, and A; all of the minimal model parameters. Thus the constancy of the relationship between $LG(fc)$ and R during both wakefulness and sleep can provide support for the appropriateness of the model formulation for LG. We have studied 21 healthy adolescents and adults (ages 9 – 38 yr) during sleep. Initial measurements were made as described under wakefulness, but in the supine posture. Additional instrumentation included; EEG (C4–A1), bitemporal EOG, and submental EMG allowing sleep staging by the criterion of Rechtschaffen and Kales[15], as well as oxygen saturation (Nellcor N–100). The plexiglass hood was modified for supine operation. After sleep onset, FIO_2 was cycled between .21, .15, and .12 with approximately 45 minutes at each level following a 10 minute equilibration period. As during wakefulness, at each FIO_2, $LG(fc)$ was estimated using a step–increase in $FICO_2$ and R was estimated using impulse–increases in $FICO_2$. Periods of rapid–eye–movement (REM) sleep were infrequent and were excluded from analysis. Epochs of periodic breathing were identified and analyzed as described under wakefulness.

Studies of clinical pathology

Familial dysautonomia (FD) (Riley–Day syndrome) is a generalized autonomic dysfunction which can be associated with significant sleep disordered breathing including periodic apneas[16]. In addition, significant abnormalities commonly exist in all minimal model parameters[16,17]. In contrast, congenital central hypoventilation syndrome (CCHS) is characterized by a virtual absence of both hypercapnic and hypoxic ventilatory responses. Based on model predictions, such patients would be expected to demonstrate metronomic respirations with little or no periodicity. To test the ability of the model to account for the observed ventilatory patterns under conditions of such widely ranging parameters, we studied 6 patients with FD during sleep and 2 patients with CCHS during wakefulness. The apparatus and manipulations were the same as described for healthy subjects with two exceptions: $FIO_2 = .12$ was not used for FD patients and a T–piece was fitted to the tracheostomy tube rather than using the hood for CCHS patients.

Results

Figure 4 presents the relationship between R and $LG(fc)$ over all 3 inspired oxygen levels for the 15 healthy awake men. Straight line segments connect different FIO_2s for individual subjects. The heavy solid curve is the relationship predicted by the minimal model. Notice that for individual subjects and for the group as a whole increases in LG tend to be correlated with increases in R, implying a greater tendency toward ventilatory oscillations at higher loop gains. Each point represents the mean of 1 to 5 estimates of R and the mean of 1 to 3 estimates of $LG(fc)$. Statistically, a significant ($p < .01$) correlation existed between R and $LG(fc)$ for the 15 subjects. When the model prediction for R was subtracted from each observed value, the resulting mean residual and the serial correlation of residuals were not significantly different from zero ($p > .1$ for each). Thus, the minimal model accounted for the deterministic portion of the observed relationship between R and $LG(fc)$.

Table 1 describes the 13 epochs of PB which occurred in these 15 subjects. All mathematical models of respiratory control predict that PB ensues when $LG(fc)$ exceeds unity; a prediction deriving directly from the Nyquist stability criterion[19]. The mean value of $LG(fc)$ corresponding to PB in awake healthy subjects was 1.17 with a range of $.91 - 1.65$. These epochs represented 12 of the 13 highest loop gains observed, and in no case was $LG(fc) > 1$ when PB was not observed. Thus, the minimal model formulation for $LG(f)$ appears to predict the threshold for PB with reasonable accuracy.

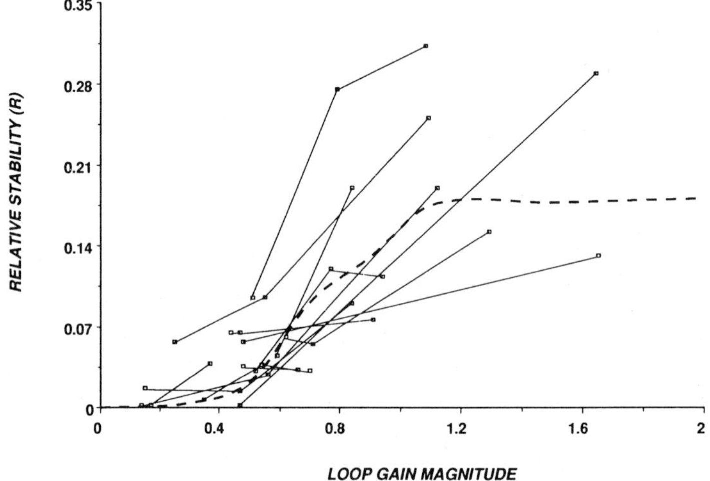

Figure 4. Comparison of observed and predicted R for 15 healthy awake men. Line segments connect FIO2s within each subject.

Table 1. Hypoxia induced periodic breathing in awake men

	# Cycles	$|LG(fc)|$	T_o	T_p
Mean	12.2	1.17*	29.0 s	36.7s**
SD	2.2	.31	8.2 s	11.5 s

Numbers of epochs of $PB = 13$. T_o, observed cycle period. T_p, predicted cycle period. *Significantly greater than unity ($p < .05$, one–tailed t–test). ** not significantly from T_o ($p > .10$, paired t–test).

Figure 5 depicts the relationship between R and $LG(fc)$ for the 21 healthy subjects during NREM sleep. Circles represent values of R derived from impulse CO_2 challenges and the squares represent the strength of ventilatory oscillations for epochs of PB (no CO_2 stimulus was applied). Notice that in no case was an underdamped impulse response observed. Every impulse CO_2 challenge during sleep resulted in an overdamped response. Thus the effect of hypoxia during sleep appeared to have a threshold behavior; if $LG(fc)$ was increased, it was increased sufficiently to create instability with concommittant PB.

When the data for the first 18 sleeping subjects were fully evaluated and this behavior became clear, 3 additional studies were performed. In these studies, FIO_2 was decreased in decrements of .01 from .15 until PB was obtained. Despite this finely graded hypoxia, no region of underdamped behavior was elicited in any of these 3 subjects (whose data are included in figure 5). In each case a threshold occurred where a .01 decrement in FIO_2 was accompanied by a transition from overdamped ventilation to unstable ventilation. As in wakefulness, PB occurred whenever $LG(fc)$ exceeded unity (figure 5).

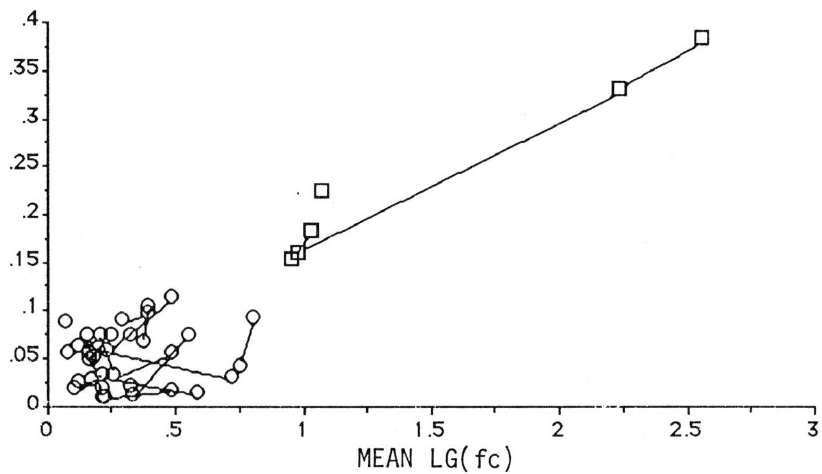

Figure 5. R vs. $LG(fc)$ for 21 normals during NREM sleep.

Table 2. Baseline parameter values – 8 FD patients, 16 controls

	$A(fc)$ (1/min/Torr)	$PvCO_2$ (Torr)	mlv (1)	D (s)	$\dot{Q}_C \cdot Ks1$ (1/min/Torr)
Mean	.18	55.5	1.96	9.1	.017
SE	.06	2.2	.19	.7	.005
Range	.04–.51	48–65	1.24–2.61	5.2–11.2	.003–.053
Control Range	.08–.35	46–56	1.11–3.76	3.6–11.0	.007–.022

Control range reflects the extreme values among 16 age and sex matched control subjects.

Figure 6 demonstrates that FD patients exhibit a relationship between R and $LG(fc)$ equivalent to healthy age and sex matched controls during sleep. Again, this relationship was accounted for by the model prediction curve ($p > .1$ for mean residual and serial correlation of residuals). Table 2 illustrates that values significantly outside the normal range occurred for every model parameter in the FD study population. Despite this fact, the model expression for $LG(f)$ would appear to adequately represent the CO_2 feedback dynamics in these patients. This interpretation is further supported by the observation that the $LG(fc)$ threshold for PB is again quantitatively accurate.

Table 3 outlines the observed values for $A(fc)$, $LG(fc)$, and R for the 2 CCHS patients. As expected, the mean $A(fc)$ and therefore $LG(fc)$ were not significantly different from zero ($p > .1$). As the model behavior predicts, the mean R is also not significantly greater than zero ($p > .1$).

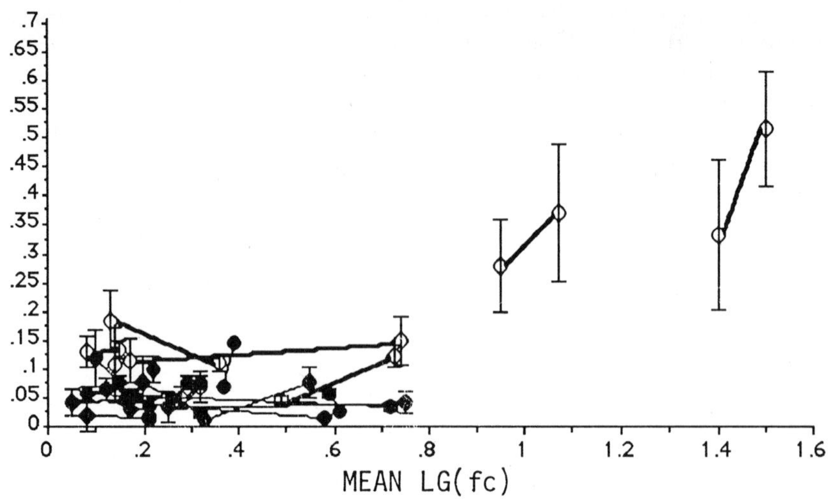

Figure 6. R vs. $LG(fc)$ for 8 FD patients (open circles) and 16 age and sex–matched normals (solid circles). Two patients and no controls displayed PB (LG(fc)> .95).

DISCUSSION

These data indicate that a linear proportional controller model of respiratory regulation can account for the relative stability of human respiration, including the conditions required for sustained periodic breathing (instability). When constrained by parameter values ex-

Table 3. Pooled results for 2 CCIIS patients

FIO_2	$PACO_2$	$|LG(fc)|$	R
.21	38.5 ± .5 Torr	.01 ± .05	−.002 ± .012
.15	38.4 ± .4 Torr	.09 ± .05	−.008 ± .009
.12	38.2 ± .7 Torr	−.13 ± .05	−.004 ± .002

Numerical values indicate mean ± SE.

perimentally evaluated in individual subjects the minimal model quantitatively predicts the magnitude of spontaneous and induced ventilatory oscillations, including PB, for that subject. These data also imply the validity of the model expressions for fc and $LG(fc)$ over a range of conditions including wakefulness, sleep, hypoxia health, generalized autonomic dysfunction, absent controller responses, and combinations of the above. Despite the agreement between model predictions and observed respiratory behavior under this wide range of conditions, a number of model limitations must be re–emphasized.

Due to the linearization performed to obtain analytic expressions for fc and $LG(f)$, the minimal model must be regarded as a "small signal" model. Thus, the model behavior may be expected to accord closely with human respiration only when $PaCO_2$ fluctuations are $\leq$ approximately 5 Torr[10]. In addition, the model becomes inadequate if mlv, $PvCO_2$, D, or $\dot{Q}_C$ fluctuate significantly[10,12]. If these conditions are satisfied, the theory of Liapunov[20] indicates that the linearized model accurately reflects the local stability of the underlying nonlinear model system about any given operating point.

Because significant fluctuations in mlv[21] as well as D and $\dot{Q}_C$ (secondary to heart rate fluctuations) may be associated with periodic breathing and periodic sleep apneas and because significant $PaCO_2$ fluctuations can occur during PB, the minimal model is not appropriate for predicting the morphology of the PB limit cycle. Similarly, the model is not appropriate for predicting the ventilatory pattern during transitions from one equilibrium condition to another. Despite these limitations, however, we found no statistically significant difference between model predictions for fc and the observed frequencies of PB in any subject group. The is consistent with the expectation that the model should correctly predict the duration of the first few cycles of PB (before the signal fluctuations significantly exceed the range of linear system operation) and the fact that no significant change in PB cycle duration was observed from the beginning to the end of any PB epoch.

Another limitation arises from the fact that the minimal model does not incorporate an explicit hypoxic response, or controller. However, because the steady–state effect of alveolar hypoxia may be appropriately characterized as an increase in the slope of the CO_2 response slope[22], the model is adequate to describe conditions of graded steady–state hypoxia, as described above.

Despite these two major limitations, this and other minimal models (cf. 11) may prove useful in evaluating the etiology of periodic respiration in conditions such as sleep onset, occlusive sleep apnea, prematurity, among others. A valuable future direction will be to incorporate the influence of changing state of consciousness into the minimal model structure. This will be especially important for the study of conditions in which both state of consciousness and ventilation vary periodically, such as sleep onset periodic breathing and occlusive sleep apnea syndrome.

ACKNOWLEDGEMENTS

This work was supported in part by the Whitaker Foundation.

REFERENCES

1. A.R. Dowell, C.E. Buckley, R. Cohen, R.E. Whalen and H.O. Sieker, Cheyne–Stokes respiration: a review of clinical manifestations and critique of physiological mechanisms, <u>Arch. Intern. Med.</u> <u>127</u>:712–726 (1971).

2. D.C. Shannon, D.W. Carley and D.H. Kelly, Periodic breathing: quantitative analysis and clinical description, <u>Pediatric Pulmonology.</u> <u>4</u>:98–102 (1988).

3. O. Klein, Untersuchungen uber das Cheyne–Stokesche atmungsphanomen, <u>Deut. Ges. Inn. Med.</u> <u>42</u>:217–222 (1930).

4. J.D. Horgan and D.L. Lange, Analog computer studies of periodic breathing, <u>IRE Trans. Biomed. Elec.</u> <u>9</u>:221–228 (1962).

5. H. Milhorn and A.C. Guyton, An analog computer analysis of Cheyne–Stokes breathing, J. Appl. Physiol. $\underline{20}$:328–333 (1965).

6. G.S. Longobardo, N.S. Cherniack and A.P. Fishman, Cheyne–Stokes breathing produced by a model of the human respiratory system, J. Appl. Physiol. $\underline{21}$:1839–1846 (1966).

7. F.S. Grodins, J. Buell and A. Bart, A mathematical analysis and digital simulation of the respiratory control system, J. Appl. Physiol. $\underline{22}$:260–276 (1967).

8. S.T. Nugent and J.P. Finley, Periodic breathing in infants – a model study, IEEE Trans. Biomed. Eng. $\underline{34}$:482–485 (1987).

9. M.C.K. Khoo, R.E. Kronauer, K.P. Strohl, and A.S. Slutsky, Factors inducing periodic breathing in humans: a general model, J. Appl. Physiol. $\underline{53}$:644–659 (1982).

10. D.W. Carley and D.C. Shannon, A minimal model of human periodic breathing, J. Appl. Physiol. $\underline{65}$:1400–1409 (1988).

11. A. elHefnawy, G.M. Saidel and E.N. Bruce, CO_2 control of the respiratory system: plant dynamics and stability analysis, Ann. Biomed. Eng. $\underline{16}$(5):445–61 (1988).

12. D.W. Carley and D.C. Shannon, Relative stability of human respiration during progressive hypoxia, J. Appl. Physiol. $\underline{65}$:1389–1399 (1988).

13. H. Watson, The technology of respiratory inductive plethysmography. in: "Third International Symposium on Ambulatory Monitoring," Horrow, Middlesex, UK: Clinical Research Center (1979).

14. C. Dubois, Changes in alveolar PCO_2 during the breathing cycle, J. Appl. Physiol. $\underline{4}$:535–548 (1952).

15. A. Rechtschaffen and A. Kales, "A manual of standardized terminology, techniques and coring system for sleep stages of human subjects," Brain Research Institute, University of California at Los Angeles, Los Angeles (1968).

16. C. Guilleminault, J. Briskin, M. Greenfield, and R. Silvestri, The impact of autonomic nervous system dysfunction during sleep, Sleep $\underline{4}$:262–278 (1981).

17. J. Dancis, Familial dysautonomia (Riley–Day syndrome), in: "Autonomic Failure. A Textbook of Clinical Disorders of the Autonomic Nervous System," R. Banister (Ed), Oxford, UK: Oxford University Press (1984).

18. J. Filler, A. Smith, S. Stone and J. Dancis, Respiratory control in familial dysautonomia, J. Pediat. $\underline{66}$:509–516 (1965).

19. R. C. Dorf, "Modern Control Systems," Menlo Park, CA: Addison–Wesley (1974).

20. M. Minorsky, "Introduction to Nonlinear Mechanics," Ann Arbor, MI: Edwards (1947).

21. R. M. Aronson, C.G. Alex, E. Önal, and M. Lopata, Changes in end–expiratory volume during sleep in patients with occlusive apnea, J. Appl. Physiol. $\underline{63}$:1642–1647 (1987).

22. N.H. Edelman, P.E. Epstein, S. Lahiri and N.S. Cherniack, Ventilatory responses to transient hypoxia and hypercapnia in man, Respir. Physiol. $\underline{17}$:302–314 (1973).

SLEEP STATE AND PERIODIC RESPIRATION

Allan I. Pack,[1] Allan Gottschalk,[2] Michael Cola,[1] Adrian Goldszmidt[1]

[1] Pulmonary Section, Department of Medicine and [2] Department of Anesthesia
University of Pennsylvania, Philadelphia, PA

Periodic breathing occurs commonly in stage 1-2 non-rapid-eye-movement (NREM) sleep.[1–4] When pronounced, apnea may occur at the nadir of ventilation. Thus, sleep apnea may be considered as a severe form of periodic breathing (see discussion, ref. 5). The degree of periodic breathing, however, varies with sleep stage. This was recognized in early, relatively qualitative studies.[1,2] Recent quantitative studies in our laboratory have both confirmed and extended such earlier observations.[4] The results of these studies led us to the hypothesis that oscillations in ventilation were secondary to oscillations in sleep state. Thus, in addition to the experimental studies, we have performed a theoretical study to determine the implications of the coupling between sleep state and ventilatory drive.

PERIODIC BREATHING AND SLEEP STATE: INITIAL EXPERIMENTAL OBSERVATIONS

We studied eight elderly subjects with periodic breathing during stage 1-2 sleep and seven elderly subjects with apneas in this stage (5 subjects with primarily obstructive apneas, 2 with central). In the elderly, periodic ventilation and apneas are particularly common during this stage of sleep.[3,6] Ventilation was measured with a respiratory inductance plethysmograph, which has the advantage of not interfering with ventilatory pattern. Periodicity was examined using digital comb filtering. The application of this spectral analysis technique to the study of respiration was described originally by Brusil et al[7] and has been employed extensively in studies of periodic respiration in the neonatal period.[8,9] One advantage of this technique is that it allows examination of the temporal variation in the degree of periodicity. In our application of the technique, the filters employed were one-half of an octave wide and their center frequencies were one-third of an octave apart. The center frequencies of the filters we employed were such that their periods were 20, 24, 28, 36, 44, 56, 68, 88 and 112 s/cycle. The average power of each filter output was calculated. To characterize the magnitude of periodicity in ventilation, we determined the filter that had the maximum average power and noted this peak power. For further description of this analysis approach, see reference 4.

In the elderly subjects studied, we found the most marked periodicities of ventilation in stage 1-2 sleep (see Fig. 1). Periodicities were also present in wakefulness. Their magnitude in wakefulness was less than in stage 1-2 sleep (see Fig. 1). Interestingly, there was a relationship between the degree of periodic breathing in individual subjects in stage 1-2 sleep and in wakefulness. Subjects who exhibited apneas in sleep had more marked periodic breathing during wakefulness than subjects without apneas.

Modeling and Parameter Estimation in Respiratory Control
Edited by M.C.K. Khoo
Plenum Press, New York

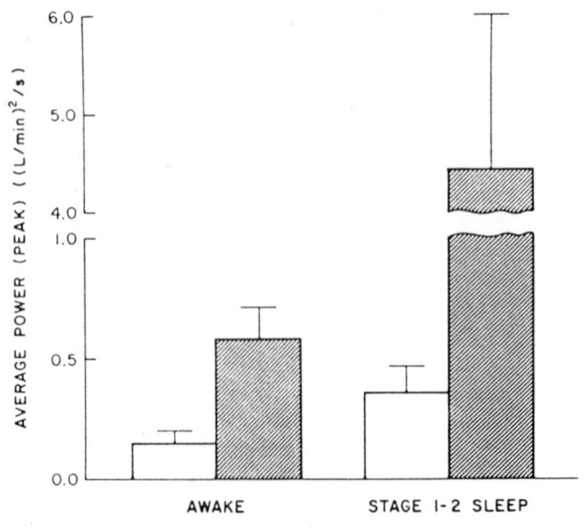

Figure 1

Mean of the peak of average power of the ventilatory oscillation. For each subject we determined the filter with the greatest average power (peak power). The mean and standard error of the mean were calculated for eight subjects without apnea (☐) and seven subjects with apnea (▨). (Reproduced by permission of the Journal of Applied Physiology (4).)

Ventilation in these subjects was most stable in stage 3-4 sleep. Sufficient data in this stage of sleep were obtained in only 4 subjects without apnea and 2 subjects with apnea. In each subject (see Fig. 2), ventilatory oscillations were virtually absent in stage 3-4 sleep.

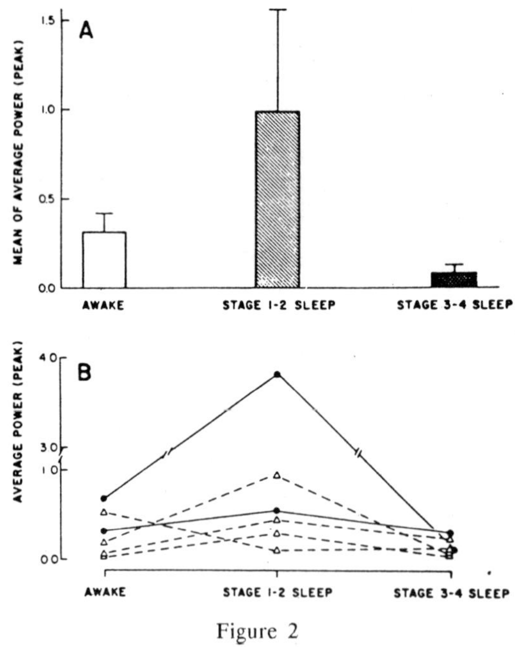

Figure 2

Peak of the average power in wakefulness, stage 1-2 sleep and stage 3-4 sleep in 4 subjects without apnea (Δ- - -Δ) and 2 subjects with apnea (●——●). The top panel (A) shows the mean ± SE of peak average power over all six subjects. The lower panel (B) shows data in individual subjects. (Reproduced by permission from the Journal of Applied Physiology (4).)

These data indicate that within an individual subject there is a marked variation in the degree of periodic breathing in different states. Periodic breathing in the elderly subjects we studied is present in wakefulness, most marked in stage 1-2 NREM sleep, and virtually absent in stage 3-4 NREM sleep. The mechanisms which account for these differences are not well established.

PERIODIC BREATHING DURING HYPOXIA

One theory is that periodic breathing is produced by unstable operation of the peripheral and central chemoreceptors.[5,10] Based on this theory, the critical variables which determine whether periodic breathing occurs are the ventilatory response to alterations in blood gas tensions and the circulatory delay between the lung and the chemoreceptors. Since hypoxia increases the gain of the peripheral chemoreflex, one would predict that hypoxia will result in periodic breathing. Hypoxia does produce periodic breathing during sleep as shown by studies at high altitude[11,12] and during breathing of hypoxic gas mixtures.[13] A theoretical analysis of the chemical feedback system[10] predicts that the cycle time of the ventilatory oscillations will be of the order of 20-30 s and that with increasing hypoxia the cycle time of the oscillations will decrease. Both of these predictions agree with experimental findings.[14] Moreover, subjects with lower ventilatory responses are less likely to develop periodic breathing during hypoxia.[11,12]

Although the experimental observations discussed above, usually made in stage 1-2 sleep, are in agreement with this theory, our previous data on the effect of sleep state on the normally occurring oscillations in ventilation led us to question what effect sleep state has on the oscillations in ventilation that are produced by hypoxia. To address this, we studied healthy young adults in whom hypoxia was produced by inspiration of a gas containing 12% oxygen. Measurements of ventilation were made using respiratory inductance plethysmography. In these studies, we observed that even with hypoxia ventilation was much less periodic in stage 3-4 sleep (see Fig. 3).

STAGE 1-2 SLEEP

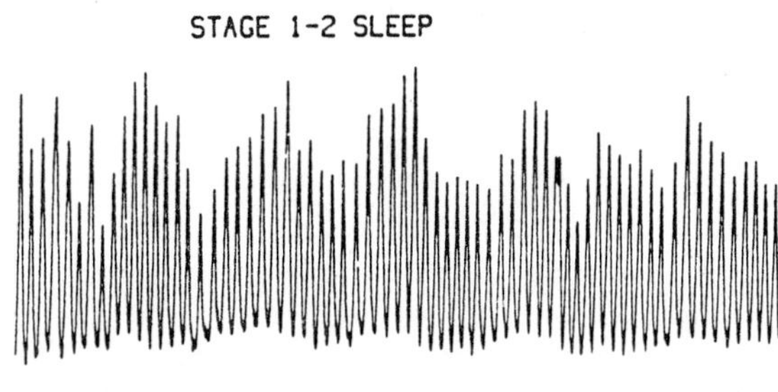

CHEST WALL
MOTION

STAGE 3-4 SLEEP

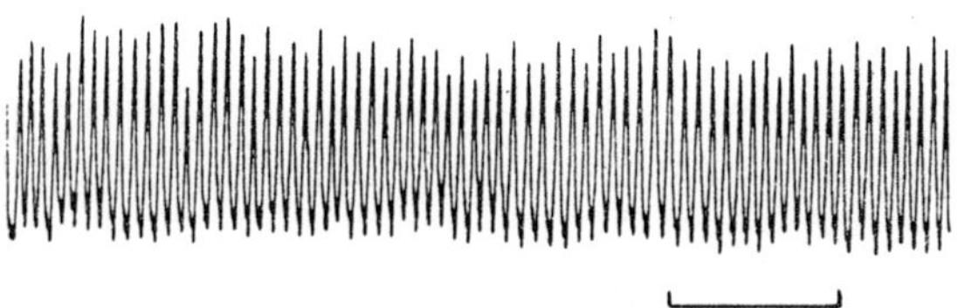

Figure 3

Chest wall motion in a single young subject, as measured by a respiratory inductance plethysmograph, during breathing of a hypoxic gas mixture (12% oxygen) in stage 1-2 sleep (top panel) and stage 3-4 sleep (lower panels). That the degree of periodicity is more marked in stage 1-2 sleep is evident.

These data add weight to the concept that the stage of sleep is important in determining the degree of periodic breathing. The explicit role of sleep state is not considered in previous theoretical analyses of mechanisms producing periodic breathing during sleep.[5,10]

A MODEL FOR PERIODIC RESPIRATION INCORPORATING SLEEP STATE AS AN EXPLICIT VARIABLE

To address this deficiency, we constructed a model which contains sleep state as an explicit variable. The model is an elaboration of that originally developed by Khoo et al.[10] The modifications include: 1) replacement of the equations describing the carbon dioxide and oxygen saturation curves by those described by Kelman;[15,16] b) a new set of controller equations which include state as an explicit parameter; c) incorporation of an arousal function; and d) allowing airway obstruction to occur. These modifications will now be described in greater detail.

We formulated the chemical drive to ventilation ($\dot{V}_{CHEM}$) as the sum of peripheral ($\dot{V}_P$) and central components ($\dot{V}_C$),

$$\dot{V}_{CHEM} = \dot{V}_P + \dot{V}_C. \tag{1}$$

Total ventilatory drive ($\dot{V}_E$) depends on the chemical drive to ventilation ($\dot{V}_{CHEM}$), a drive coupled to state (RAS) and a threshold (TH);

$$\dot{V}_E = [G\ \dot{V}_{CHEM} + RAS - TH], \tag{2}$$

where G is the overall gain of the chemical drive to ventilation.

$$\text{and } [x] \begin{cases} x & x>0 \\ 0 & x \geq 0 \end{cases}. \tag{3}$$

Our peripheral chemical drive ($\dot{V}_P$) is given by

$$\dot{V}_P = \alpha \left[\frac{PaCO_2 - 13}{PaO_2 - 20} \right] \tag{4}$$

where $PaCO_2$ and PaO_2 are respectively the arterial carbon dioxide and oxygen tensions at the peripheral chemoreceptor, and α is a constant determining the relative contribution of the peripheral drive to the total chemical drive to ventilation. Equation [4] was obtained by fitting a hyperbola to the data obtained by Lahiri and DeLaney,[17] who examined the effect of variations in pO_2 and pCO_2 on carotid body activity in single fiber studies in cats. The drive from the central chemoreceptor in the steady state is given by

$$\dot{V}_C = G_C\ PaCO_2 \tag{5}$$

where G_C is the slope of the carbon dioxide response curve and $PaCO_2$ is the arterial carbon dioxide tension in the circulation to the medulla. We did not include a threshold here because carbon dioxide tensions as low as 4 torr can influence ventilation when the majority of the ventilatory drive comes from the peripheral chemoreceptors.[18]

In addition to incorporating a state term explicitly in the controller equation, we included an arousal function. Arousal can be produced by both hypoxia[19] and hypercapnia.[20] Recent studies, reported only in abstract form,[21] indicate that esophageal pressure at arousal is similar regardless of the combination of hypercarbia and hypoxia that produces the arousal. We have, therefore, viewed arousal as being related to the total chemical drive to ventilation ($\dot{V}_{CHEM}$ in Eq. [1]). Once an obstructive apnea occurred, an arousal is envisaged to take place when $\dot{V}_{CHEM}$ is greater than a fixed threshold. In determining this threshold we have utilized data on arousal thresholds in studies with eucapnic hypoxia[19] and hyperoxic hypercapnia.[20] We obtained the arousal threshold by substituting into Eq. [1] as follows

$$\dot{V}_{CHEM}\ (PaO_2^{AR}, PaCO_2^{AR}) = \text{Arousal Threshold}, \tag{6}$$

where PAO_2^{AR} and $PaCO_2^{AR}$ are the arterial oxygen and carbon dioxide tensions at arousal.

A simple function governs airway obstruction in our model. We envisage that the airway will close down whenever an apnea occurs, i.e. whenever $\dot{V}_E$ falls to zero, and the airway will reopen only when an arousal takes place.

To model the changes that take place with sleep, we need to determine the magnitude of the change in the state dependent input to the ventilatory controller (RAS) at sleep onset. We begin this analysis by considering the change that takes place with sleep in metabolic CO_2 production. We consider that CO_2 production while asleep is proportional to that in the awake state;

$$B \; \dot{V}CO_2(\text{Awake}) = \dot{V}CO_2(\text{Sleep}), \qquad [7]$$

where B is a constant ≤ 1. The constant B accounts for the reduced metabolic rate during sleep. Here we use a value of 0.8 for this based on experimental data.[22] Since $\dot{V}CO_2 = c\dot{V}_A PaCO_2$ where c is a constant and $\dot{V}CO_2$ is the rate of CO_2 production,

$$B \; \dot{V}_A^{AW} \; Pa^{AW}CO_2 = \dot{V}_A^{S} \; Pa^{S}CO_2, \qquad [8]$$

where the superscripts AW and S refer to wakefulness and sleep respectively. If we simplify things by making the approximations that $\dot{V}_E = \dot{V}_A$ and $B = 1$, we may then write

$$\dot{V}_E^{AW} \; Pa^{AW}CO_2 = \dot{V}_E^{S} \; Pa^{S}CO_2. \qquad [9]$$

If we then substitute Eq. [2] into [9], we obtain

$$(G^{AW}(\dot{V}_p^{AW} + \dot{V}_c^{AW}) + k^{AW}) \; Pa^{AW}CO_2 = (G^{S}(\dot{V}_p^{S} + \dot{V}_c^{S}) + k^{S}) \; Pa^{S}CO_2 \qquad [10]$$

where we have let k = RAS - TH. For our approximate analysis, let us begin with the simple case where $G^{AW} = G^{S} = 1$, the peripheral and central drives are the same function of $PaCO_2$, and the PaO_2 is not significantly altered with changes in ventilation. If $k^{S} = k^{AW} + \Delta k$ (where Δk is the change in the RAS - TH at at sleep onset) and $Pa^{S}CO_2 = Pa^{AW}CO_2 + \Delta Paco_2$ (where $\Delta PaCO_2$ is the change in PCO_2 with sleep), consider how $Pa^{S}CO_2$ varies as a function of a change in k. This can be affected by changes in either the state dependent input (RAS) or the ventilatory threshold (TH). This results in an effectively linear relationship between the change in PCO_2 with sleep (ΔPCO_2) and the change in the variable RAS - TH, i.e., Δk, over the physiological range.

Since we know that the change in PCO_2 with sleep is, in man, of the order of 2-5 torr,[22] we can obtain from Eq. [10] the change in k with sleep which would result in this change in pCO_2. With Δk now fixed, we can next examine the resulting change in pCO_2 which will occur with sleep for different values of the gain of the chemical ventilatory control system (G in Eq. [2]). For a fixed sleep state change, there is a hyperbolic relationship between the change in pCO_2 with sleep and the gain of the ventilatory control system (see Fig. 4). This hyperbolic relationship resembles that found experimentally by Gothe et al.[23] If the equations are modified to permit the PaO_2 to vary with ventilation by use of the alveolar gas equation, we find no significant deviation from the hyperbolic relationship depicted in Figure 4. The foregoing analysis adds validity to our approach for coupling the sleep state with the ventilatory drive. In addition, we have obtained the means of parameterizing this aspect of the model.

We are currently studying the dynamics of the entire model system using computer simulation. One early result from this approach is the influence of the rate of change in the sleep state at sleep onset on the emergence of significant periodicities in ventilation. In these simulations we allow the transition to sleep to take place at different rates. When the sleep process begins, k^{AW} is changed by Δk to k^{S} linearly over a specified transition time (TT). If the conditions in Eq. [6] are met, an arousal occurs and k^{S} is immediately changed to k^{AW}. Once the values of pCO_2 and pO_2 fall below the arousal threshold the transition to the sleep state takes place as before. With a fixed value of gain of the chemical ventilatory control system (G = 1), we find that the events at sleep onset are critically dependent on the transition time (see Fig. 5).

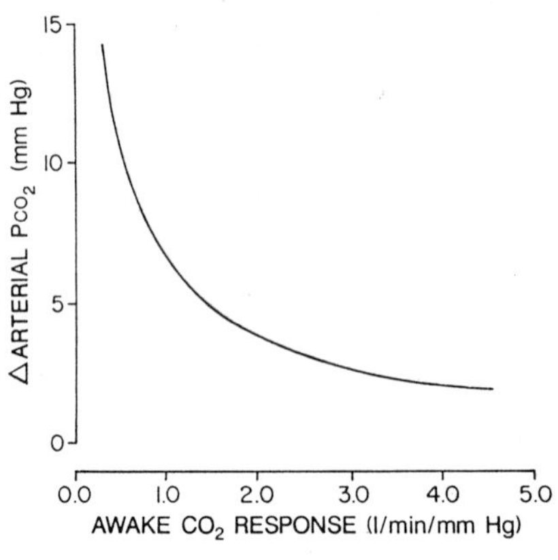

Figure 4

Predicted dependence of the change in $PaCO_2$ with sleep and the awake ventilatory response to CO_2. This prediction agrees well with the experimental observations of Gothe et al.[23]

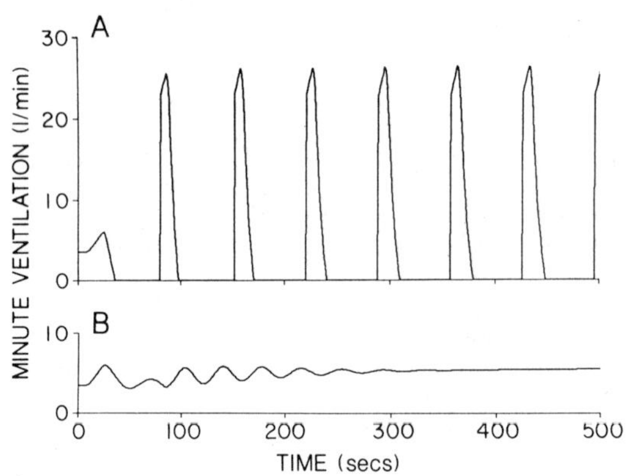

Figure 5

Predicted ventilation at sleep onset with two different values of transition time (A) 15 s (top panel) and (B) 60 s (lower panel). In each case we used the same absolute magnitude of the change in state dependent input to the ventilatory control system at sleep onset and the other parameters were identical (for further discussion, see text).

With a rapid transition to sleep (15 s), repetitive apneas occur (see top panel, Fig. 5). However, with a longer transition time (60 s), there are no apneas. We believe that at sleep onset the change in ventilation produced by the fall in the RAS term in Eq. [2] is minimized by the compensatory action of the chemoreceptors responding to the resulting hypoxia and hypercapnia. With rapid transitions there is inadequate time for such compensation so that apneas occur. With this argument we would postulate that low values of chemoreceptor gain would augment this effect of state transition, since the magnitude of the compensation would be less. This is shown by simulation (see Fig. 6). With a lower value of gain (G = 0.25), apneas and periodicities occur with both fast (15 s) and slow (60 s) transitions. This effect of low gain may explain the clinical phenomena that apneas occur commonly in patients with hypothyroidism,[24,25] who are known to have low ventilatory responses,[26] and in the elderly,[6] who also have reduced ventilatory responses to chemical stimuli.[27,28]

These data imply that a rapid transition to sleep is more likely to be associated with apnea. Since sleepiness should result in a more rapid transition from wakefulness to sleep, there is conceivably a vicious cycle in patients with the sleep apnea syndrome. Since the disease leads to excessive daytime sleepiness, such sleepiness may potentiate the tendency for the development of apnea.

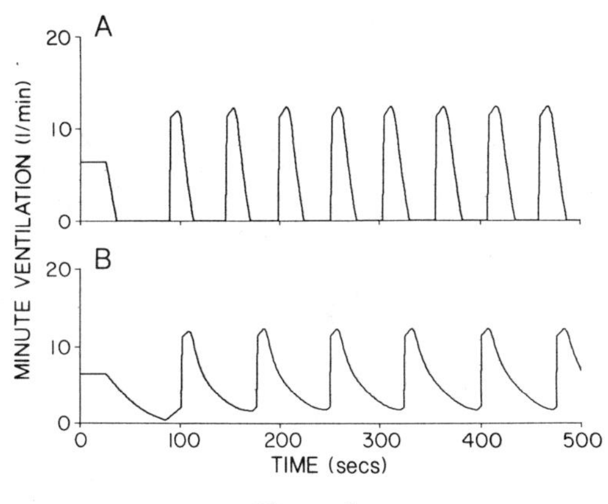

Figure 6

Predicted ventilation at sleep onset with the two values of transition time shown in Figure 5: (A) 15 s (top panel) and (B) 60 s (middle panel). The parameter values and change in k at sleep onset are identical to the simulations performed for Figure 5, except for the low value of chemoreceptor gain (G = 0.25). With a lower value of gain, periodicities occur even at longer transition times (compare with Fig. 5).

RELATIONSHIP BETWEEN OSCILLATIONS IN VENTILATION AND OSCILLATIONS IN STATE

Our theoretical results suggest a tight coupling between sleep state and ventilation, and that this coupling can lead to periodic behavior and apneas. During the phase of the cycle before apnea when ventilation is falling, there is a progressive decline in the state-dependent input. Apnea is terminated by arousal and a sudden increase in this state-dependent effect. While these theoretical predictions seem plausible, they are difficult to prove since the state-dependent input is not directly measurable. It is likely to be related to the firing of cells within the brainstem reticular formation.[29] Although this important variable cannot be directly measured, there are potentially indirect approaches to its assessment. Changes in state are associated with change in several variables. One such variable is the electroencephalogram, changes in which are of primary importance in current definitions of sleep state.[30] One method to quantify such changes is to examine changes in the frequency content of the EEG, since sleep is associated with a progressive slowing of its major frequency component.[31,32] In experimental studies we have, therefore, examined the relationship between the oscillatory changes in ventilation and the oscillatory changes in EEG frequency content. These studies were also done in elderly subjects in whom periodic breathing and apneas during stage 1-2 sleep are particularly common.[3,6] Ventilation was measured by a respiratory inductance plethysmograph and we recorded the activity of the EEG from the P_3O_4 electrode. We divided the recording into 5.12 s epochs and for each epoch computed the power spectrum of the data. From each epoch we extracted two measures: a) alpha power, i.e. the power in the band from 8-12 Hz and b) mean frequency. The latter was obtained by examining the spectrum between 2 and 25 Hz. (We chose 2 Hz as the lower limit to avoid low frequency components related to movement artifact.) Mean frequency is defined as:

$$Mf = \frac{\sum_{2}^{25} P(fi)/fi}{\sum_{2}^{25} P(fi)} \qquad [11]$$

where fi is the frequency in increments of 0.195 Hz (the spectral resolution) and P(fi) is the power spectral density calculated from 2 to 25 Hz using the fast fourier transform.

This resulted in a time series of both alpha power and mean frequency of the EEG, with data points being at 5.12 s intervals. To relate this to ventilation, we resampled the breath-by-breath measures of ventilation using the approach described by Wagner et al[9] to also produce a time series with measurements of ventilation at 5.12 s intervals.

In the eight elderly subjects studied, we observed oscillations in both ventilation and the variables derived from the EEG (alpha power, mean frequency) in stage 1-2 sleep (see example, Fig. 7). The oscillations appear to be synchronous.

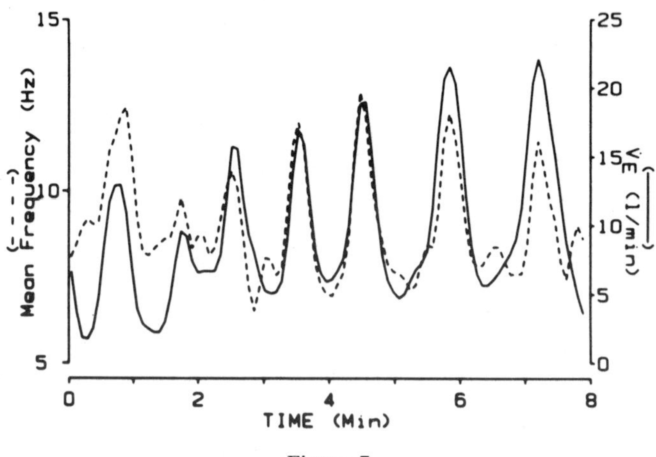

Figure 7

Changes in minute ventilation (———) (right hand axis) and mean frequency of the electroencephalogram (- - -) (left hand axis) in a single elderly subject breathing oxygen during a period of stage 1-2 sleep. There are essentially synchronous oscillations in both variables.

To examine this relationship further, we determined the coherence between the oscillations in ventilation and those in frequency content of the EEG. In all subjects we found, in stage 1-2 sleep, a high coherence: mean = 0.87 ± 0.06 (SD) for ventilation and alpha power; mean = 0.90 ± 0.04 for ventilation and EEG mean frequency. The maximum coherence was found at almost an identical cycle-time as that where we detected the maximum ventilatory oscillation using our digital comb filtering approach (see above). To further examine the temporal nature of the different oscillatory processes, we calculated the correlation coefficient between the time series for ventilation and alpha power, and the time series for ventilation and mean frequency with different values of time-lag or time-lead between ventilation and the EEG time series. In five of the eight subjects we found the maximum correlation coefficient between alpha power and ventilation at a time-lag of zero, while for mean frequency and ventilation, the maximum occurred at a time-lag of zero in six of the eight subjects. Thus, not only are the oscillations coherent, but these data suggest that they are essentially synchronous. These experimental data, albeit using indirect approaches to measurement of the state-dependent input to the ventilatory control system, support the concepts elucidated in our hypothetical model of the pathogenesis of apnea that incorporates state explicitly.

CONCLUSIONS

In conclusion, periodic breathing occurs commonly during sleep and is highly variable in the different sleep states. It is most marked in stage 1-2 sleep and virtually absent in stage 3-4 sleep. Although current theories, as to the role of instability in the chemical feedback system for respiration in the genesis of periodic respiration, probably explain periodic breathing in certain circumstances, they seem insufficient to explain all of the features of the state-related changes in periodic respiration. Theoretical analyses need to consider the important state-dependent input to the ventilatory control system in an explicit way. Indeed, initial results with this type of analysis highlight the importance of the rate of sleep onset in determining whether apneas occur or not. This critical role of changes in sleep state is supported by experimental observations which reveal synchronous oscillations in ventilation and the frequency content of the electroencephalogram. The latter is known to vary with sleep state.

ACKNOWLEDGMENTS

We are extremely grateful to Dr. Richard O. Davies for helpful discussions during these studies and Mr. Daniel C. Barrett for help in preparation of this manuscript. The studies reported here were supported in part by HL-42236 and AG-03934.

REFERENCES

1. K. Bulow, Respiration and wakefulness in man, Acta Physiol. Scand. Suppl. 209:5-110 (1963).

2. P. Webb, Periodic breathing during sleep, J. Appl. Physiol. 37:899-903 (1974).

3. E.T. Shore, R.P. Millman, D.A. Silage, D.C. Chung, and A.I. Pack, Ventilatory and arousal patterns during sleep in normal young and elderly subjects. J. Appl. Physiol. 59:1607-1615 (1985).

4. A.I. Pack, D.A. Silage, R.P. Millman, H. Knight, E.T. Shore, and D-C.C. Chung, Spectral analysis of ventilation in elderly subjects awake and asleep. J. Appl. Physiol. 64:1257-1267 (1988).

5. G.S. Longobardo, B. Gothe, M.D. Goldman, and N.S. Cherniack, Sleep apnea considered as a control system instability, Respir. Physiol. 50:311-333 (1982).

6. M.A. Carskadon and W.C. Dement, Respiration during sleep in the aged human, J. Gerontol. 36:420-423 (1981).

7. P.J. Brusil, T.B. Waggener, and R.E. Kronauer, Using a comb filter to describe time-varying biological rhythmicities, J. Appl. Physiol. 48:557-561 (1980).

8. T.B. Waggener, I.D. Frantz III, A.R. Stark, and R.E. Kronauer, Oscillatory breathing patterns leading to apneic spells in infants, J. Appl. Physiol. 52:1288-1295 (1982).

9. T.B. Waggener, A.R. Stark, B.A. Cohlan, and I.D. Frantz III, Apnea duration is related to ventilatory oscillation characteristics in newborn infants, J. Appl. Physiol. 57:536-544 (1984).

10. M.C. Khoo, R.E. Kronauer, K.P. Strohl, and A.S. Slutsky, Factors inducing periodic breathing in humans: a general model, J. Appl. Physiol. 53:644-659 (1982).

11. S. Lahiri, K. Maret, and M.G. Sherpa, Evidence of high altitude sleep apnea on ventilatory sensitivity to hypoxia, Respir. Physiol. 52:281-301 (1983).

12. D.P. White, K. Gleeson, C.K. Pickett, A.M. Rannels, A. Cymerman, and J.V. Weil, Altitude acclimatization: influence on periodic breathing and chemoresponsiveness during sleep, J. Appl. Physiol. 63:401-412 (1987).

13. A. Berssenbrugge, J. Dempsey, C. Iber, J. Skatrud, and P. Wilson, Mechanisms of hypoxia-induced periodic breathing during sleep in human, J. Physiol. (Lond.) 343: 507-526 (1983).

14. T.B. Waggener, P.J. Brusil, R.E. Kronauer, R.A. Gabel, and G.F. Inbar, Strength and cycle time of high-altitude ventilatory patterns in unacclimatized humans. J. Appl. Physiol. 56:576-581 (1984).

15. G.R. Kelman, Digital computer subroutine for the conversion of oxygen tension into saturation, J. Appl. Physiol. 21:1375-1376 (1966).

16. G.R. Kelman, Digital computer procedure for the conversion of PCO_2 into blood CO_2 content, Respir. Physiol. 3:111-115 (1967).

17. S. Lahiri and R.G. DeLaney, Stimulus interaction in the responses of carotid body chemoreceptor single afferent fibers, Respir Physiol. 24:249-266 (1975).

18. A. Berkenbosch, J.H. van Beek, C.N. Olievier, J. De Goede, and P.H. Quanjer, Central respiratory CO_2 sensitivity at extreme hypocapnia, Respir. Physiol. 55:95-102 (1984).

19. M. Berthon-Jones and C.E. Sullivan, Ventilatory and arousal responses to hypoxia in sleeping humans, Am. Rev. Respir. Dis. 125:632-639 (1982).

20. M. Berthon-Jones and C.E. Sullivan, Ventilation and arousal responses to hypercapnia in normal sleeping humans, J. Appl. Physiol. 57:59-67 (1984).

21. K. Gleeson, C.W. Zwillich, and D.P. White, Arousal from sleep in response to ventilatory stimuli occurs at a similar degree of ventilatory effort irrespective of the stimulus, Am. Rev. Respir. Dis. 139:A82 (1989).

22. G. Bowes and E.A. Phillipson, Control of breathing during sleep in: "Handbook of Physiology, Section 3, Volume II, Part 2", 1986, p. 647-689.

23. B. Gothe, M.D. Altose, M.D. Goldman, and N.S. Cherniack, Effect of quiet sleep on resting and CO_2-stimulated breathing in humans, J. Appl. Physiol. 50:724-730 (1981).

24. K.R. Rajagopal, P.H. Abbrecht, S.S. Derderian, C. Pickett, F. Hofeldt, C.J. Tellis, and C.W. Zwillich, Obstructive sleep apnea in hypothyroidism, Ann. Intern. Med. 101: 491-494 (1984).

25. R.P. Millman, J. Bevilacqua, D.D. Peterson, and A.I. Pack, Central sleep apnea in hypothyroidism, Am. Rev. Respir. Dis. 127:504-507 (1983).

26. C.W. Zwillich, D.J. Pierson, F.D. Hofeldt, E.G. Lufkin, and J.V. Weil, Ventilatory control in myxedema and hypothyroidism, N. Engl. J. Med. 292:662-665 (1975).

27. R.S. Kronenberg and C.W. Drage, Attenuation of the ventilatory and heart rate responses to hypoxia and hypercapnia with aging in normal men, J. Clin. Invest. 52:1812-1819 (1973).

28. D.D. Peterson, A.I. Pack, D.A. Silage, and A.P. Fishman, Effects of aging on ventilatory and occlusion pressure responses to hypoxia and hypercapnia, Am. Rev. Respir. Dis. 124:387-391 (1981).

29. J. Orem, I. Osorio, E. Brooks, and T. Dick, Activity of respiratory neurons during NREM sleep, J. Neurophysiol. 54:1144-1156, 1985.

30. A. Rechtschaffen and A. Kales (editors), "A Manual of Standardized Techniques and Scoring System for Sleep Stages of Human Subjects", Brain Information Service, Brain Research Institute, Los Angeles, CA (1968).

31. I. Gath and E. Bar-On, Classical sleep stages and the spectral content of the EEG signal, Int. J. Neurosci. 22:147-155, 1983.

32. Hasan, J., Differentiation of normal and disturbed sleep by automatic analysis, Acta Physiol. Scand. (Suppl.) 526:1-103, 1983.

MODELING THE EFFECT OF SLEEP–STATE ON RESPIRATORY STABILITY

Michael C.K. Khoo

Biomedical Engineering Dept., University of Southern
California, Los Angeles, CA 90089

INTRODUCTION

The recognition that sleep significantly lowers resting ventilation and alters breathing pattern dates back to studies published in the nineteenth century[1-4]. Mosso[4] noted a tendency in sleeping subjects to exhibit a periodic waxing and waning of respiration, similar to the pattern seen during Cheyne–Stokes breathing. Magnussen[5] suggested a connection between this phenomenon and the alternating periods of wakefulness and sleep that he observed. Subsequently, with the use of electroencephalography, Aserinsky and Kleitman[6] revealed the existence of rapid–eye–movement (REM) sleep and found that these periods were associated with a marked increase in respiratory variability. Bulow[7] showed that periodic breathing (PB) often occurred during drowsiness and light NREM sleep, but in the deeper stages of NREM sleep (or slow–wave sleep (SWS)), ventilation became highly regular. These findings have been confirmed by others[8,9]. However, in some reports[10,11], PB has been found to occur during REM sleep as well.

A possible explanation for the appearance of PB in light NREM sleep (stages 1 and 2) but not in SWS has been proposed in a recent theoretical study[12]. Using a computer model of respiratory control, we found that, although the ventilatory response to CO_2 is progressively reduced, system stability decreases in light NREM sleep before eventually increasing with further deepening of sleep state[12]. This suggested that PB which occurs during light sleep may be mediated by the same kind of high loop–gain instability[13] that produces PB in hypoxic subjects[14], patients with decreased cardiac output or prolonged circulation times[15], or neonates[16]. In cases with depressed controller gain, periodic ventilatory fluctuations can still occur as a result of repetitive arousal from sleep if the swings in blood gases are sufficiently large.

In this paper, we examine in greater detail the interactions between the chemical and non–chemical factors that can lead to respiratory instability during the various stages of sleep. In particular, we focus attention on how hypercapnia, hyperoxia, and the rate and amplitude of withdrawal of the non–chemical 'wakefulness' drive may affect model stability. The effects of phasic (breath–to–breath) fluctuations in ventilatory drive in different sleep stages are also studied.

MODEL DESCRIPTION

The model used in the present study has been described in detail previously[12]. In brief, the model consists of a single gas exchanging lung compartment in series with a dead space, a parallel body tissues compartment representing gas transport at the metabolic level, a peripheral chemoreflex loop responsive to arterial P_{CO_2} (P_{aCO_2}) and P_{O_2} (P_{aO_2}), a central chemoreflex loop responsive only to brain compartment P_{CO_2} (P_{bCO_2}), and a respiratory center where chemical and non–chemical drives are summed and translated into tidal volume (V_T) and breath timing components. Nonlinear blood–gas dissociation curves, the Bohr–Haldane effect, and the effect of P_{aCO_2} and P_{aO_2} on cerebral blood flow are also incorporated. A block diagram representation of the model is shown in Fig.1.

The integration of the component drives to breathe is modeled in the same way as described previously[12], except for some minor modifications. Total ventilatory drive ($\dot{V}_I \equiv V_T/T_I$) is assumed to be proportional to the sum of the combined chemoreflex drive (D) and a non–chemical drive (S):

$$\dot{V}_I = G_w[D + S]$$

where $\dot{V}_I$ becomes zero if the argument within [ ] becomes negative, and G_w is an attenuation factor which depends on sleep state. S is assumed to be composed of a tonic component (S_{tonic}) and a phasic component (S_{phasic}):

$$S = S_{tonic} + S_{phasic}$$

S_{phasic} reflects random breath–to–breath fluctuations in respiratory drive not related to the chemoreflexes. S_{tonic} is assumed to depend on sleep/waking state, which is in turn quantified in terms of the index E:

$$\begin{aligned} S_{tonic} &= -3.57\,S_0\,E &\text{for} \quad 0 \le E < 0.28 \\ &= -S_0 &\text{for} \quad 0.28 \le E \le 1 \end{aligned}$$

Note that S_{tonic} is zero in the fully awake condition but negative during sleep to represent the withdrawal of the "wakefulness" stimulus[17]. As in our previous paper, G_w depends on state in the following manner:

$$G_w = 1 - 0.4\,E$$

The following ranges of values for E represent the different sleep/waking states:

$$\begin{aligned} \text{Wakefulness} &\longrightarrow E = 0 \\ \text{Drowsiness} &\longrightarrow 0 < E < 0.28 \\ \text{NREM–1 sleep} &\longrightarrow 0.28 \le E < 0.53 \\ \text{NREM–2 sleep} &\longrightarrow 0.53 \le E < 0.89 \\ \text{NREM–3 sleep} &\longrightarrow 0.89 \le E < 0.99 \\ \text{NREM–4 \& REM} &\longrightarrow 0.99 \le E \le 1 \end{aligned}$$

During the transition from wakefulness to sleep, E increases from zero towards unity with the following form of time–dependence (with $\tau = 0$ representing the start of each transition):

$$E = 1 - \exp(-\tau/\tau_s)$$

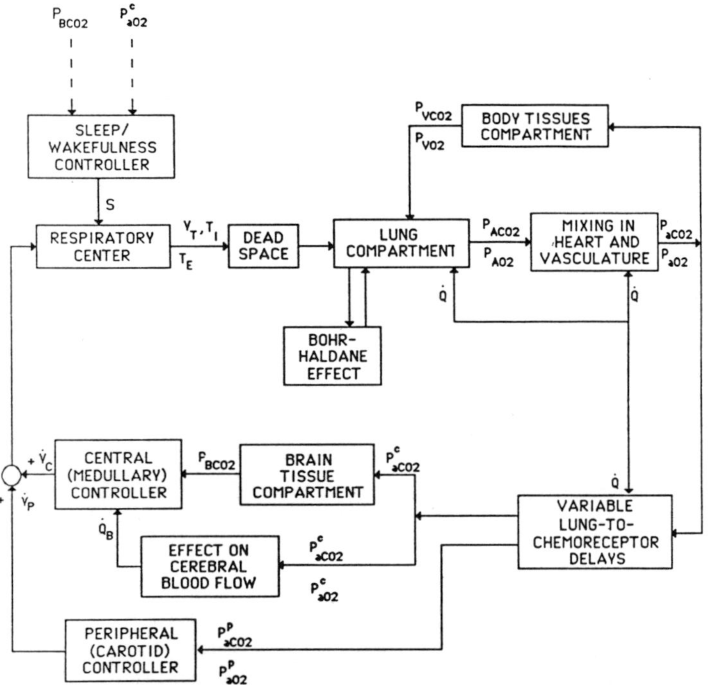

Fig. 1   Block diagram representation of the respiratory control model.

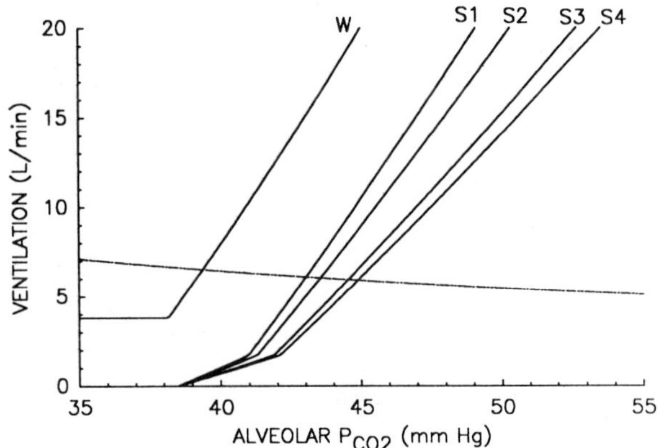

Fig. 2   Steady state ventilatory responses to CO_2 generated by the model during simulated conditions of wakefulness (W) and the four stages of NREM sleep (S1–S4). Broken curve intersecting CO_2 responses represents steady state behavior of the controlled system (lungs and body tissues).

The nominal value of the transition time constant (τ_s) is 180 s, i.e. it takes at least a minute to get into NREM-1 sleep from quiet wakefulness. However, τ_s can be changed in accordance with the rate at which sleep deepens. The magnitude of the effect of falling asleep on ventilatory drive can be altered by adjusting the value of S_0. Arousal thresholds for P_{bCO_2} and O_2 saturation are set at 55 torr and 80%, respectively, in accordance with empirical data[9]. Excursions beyond either of these limits would trigger a change in E from the current value back to zero. S_{tonic} and G_w would change correspondingly and affect the magnitude of inspiratory drive at the start of the next breath.

In wakefulness, the metabolic production rate for CO_2 (MR_{CO_2}) and consumption rate for O_2 (MR_{O_2}) are assumed to be 200 and 240 ml/min STPD, respectively. During the various stages of sleep, these rates change linearly with G_w:

$$MR_{CO_2} = 200(0.375G_w + 0.625)$$
$$MR_{O_2} = 240(0.375G_w + 0.625)$$

Unless otherwise stated, all other model equations and parameter values are the same as those published in Khoo[12].

RESULTS AND DISCUSSION

Steady state behavior

The steady state ventilatory responses to CO_2 of the model in wakefulness and the various sleep stages are shown in Fig.2. The CO_2 response slope in wakefulness is 2.54 l/min/torr, with peripheral gain making up 23% of the total chemoreflex gain. Note that the model assumes a 'dogleg' effect in the hypocapnic region (below 38 torr). With sleep stages 1, 2, 3 and 4, CO_2 sensitivity is reduced by 11%, 21%, 35%, and 40%, respectively, consistent with the observations of Bulow[7]. Also in accordance with Bulow's data, we have assumed that the rightward shift of the CO_2 response line is basically completed when stage 1 sleep is attained. The 'bend' in the lower portions of the sleep CO_2 response lines is the result of an artificial limitation (6.5 s) we have placed on the maximum expiratory duration (T_E); this is necessary because extrapolation of the empirical $V_T - T_E$ relation[12] to volumes approaching dead space produces infinitely long T_E.

The broken line intersecting the CO_2 response lines represents the steady state open-loop response of the 'plant' or controlled system portion of the model. Thus, the points of intersection represent the steady-state operating points of the various sleep/waking stages.

Stability analysis

To test the stability of the model for a given set of parameter values, the model equations are solved for the steady state, starting from arbitrary initial conditions. Subsequently, an impulsive disturbance in the form of a hyperventilatory breath (approximately 2.5 times larger than normal) is imposed and the post-sigh behavior is examined. As in our previous simulations[12], the post-sigh responses in NREM stages 1 and 2 show underdamped, oscillatory behavior, while the corresponding responses in SWS are highly damped. Our model therefore predicts that the stability of respiratory control is decreased in NREM-1 and 2, but increased as sleep deepens.

The important role of the magnitude of the wakefulness stimulus in determining system stability is shown in Fig.3. Figure 3A illustrates the post-sigh response during NREM-1 sleep in the case where S_0 assumes a value of 14.4 l/min; this is equivalent to a rightward shift of the CO_2 response line of 3.6 torr from its original position during wakefulness. Here, prior to the sigh, steady state minute ventilation ($\dot{V}_E$), alveolar P_{CO_2} (P_{ACO_2}) and arterial O_2 saturation (S_{aO_2}) are 5.9 l/min, 42.9 torr and 98.1%, respectively. These may be contrasted with the corresponding values in wakefulness of 6.5 l/min, 39.3 torr and 98.6%. However, a doubling of S_0, which is equivalent to an additional 3.1 torr shift of the CO_2 response to the right,

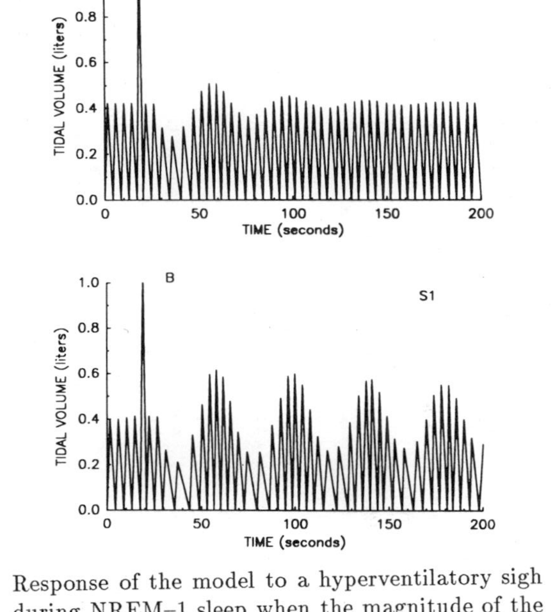

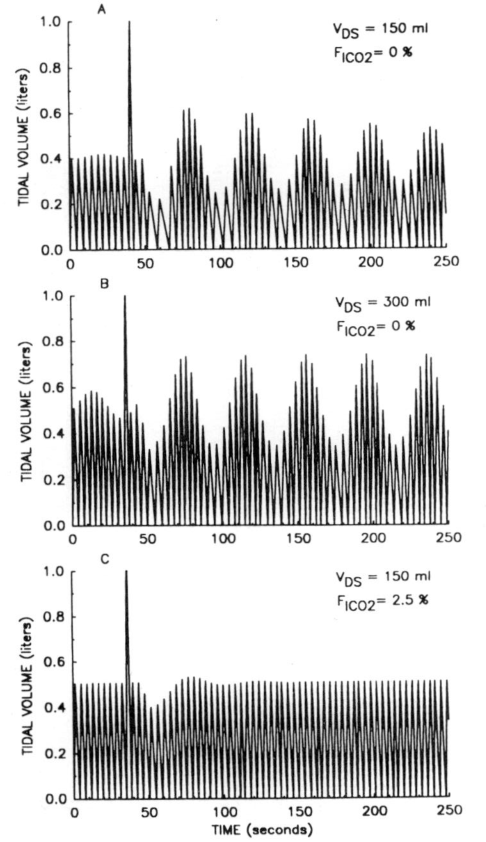

Fig. 3   Response of the model to a hyperventilatory sigh during NREM–1 sleep when the magnitude of the wakefulness stimulus (S_0) is: (A) 14.4 ℓ/ min, and (B) 28.8 ℓ/ min.

Fig. 4   Effects of hypercapnia on system stability. (A) Control condition (same post–sigh response as Fig. 3A); (B) Hypercapnia is induced by doubling dead space; (C) Hypercapnia is induced by CO_2 inhalation.

changes steady state $\dot{V}_E$, P_{ACO_2} and S_{aO_2} to 5.4 l/min, 45.9 torr and 97.5%, respectively. The dynamic post–sigh response in this case shows an almost sustained oscillation with a period of roughly 40 s (Fig.3B). Thus, the more the wakefulness stimulus contributes to ventilatory drive in the awake state, the greater will be the propensity for unstable behavior during sleep when this drive is inhibited. Furthermore, our simulations also show that, for the same S_0, the rate at which the wakefulness stimulus is withdrawn is also highly important in determining whether the onset of sleep will lead to PB; this rate is inversely proportional to τ_s. A more abrupt withdrawal of the wakefulness drive leads to a greater degree of destabilization during the early stages of sleep. These results are compatible with Bulow's observations[7] that the incidence of PB in NREM–1 and 2 is increased in cases where the level of wakefulness decreases more rapidly or in subjects who exhibit the largest rightward shifts of the CO_2 response line. Bulow also noted that the subjects who have greater CO_2 sensitivity while awake are more likely to show PB in light sleep. This further supports our belief that PB during light sleep is likely to be mediated by an underlying instability in feedback control.

How can system stability be *decreased* in light sleep but *increased* in SWS, when controller gain is progressively reduced along with the level of wakefulness? This apparent paradox can be explained quite simply if we recall that loop gain, which varies inversely with the degree of stability, is determined by both controller and plant gains. From our previous analysis[13], plant gain for CO_2 is proportional to $P_{ACO_2} - P_{ICO_2}$ but inversely proportional to alveolar ventilation ($\dot{V}_A$). With the onset of sleep, plant gain increases as a result of the elevated operating P_{ACO_2} and the decreased $\dot{V}_A$. At the same time, controller gain decreases. In light sleep, the increase in plant gain more than offsets the reduction of controller gain; as a result, loop gain is increased and the system becomes less stable. However, further decreases in controller gain overwhelm the changes in plant gain as sleep deepens; consequently, loop gain is reduced and system becomes more stable in SWS than in wakefulness.

To confirm the above conclusions, the results of additional simulations are shown in Fig.4. In the 3 cases shown, model computations are performed assuming that a steady level of NREM–2 sleep has been attained with $S_0 = 28.8$. The control case is shown in Fig.4A: a hyperventilatory sigh leads to oscillatory behavior that is finally damped out. Consider what happens when the model is made more hypercapnic through two different methods. In the first (Fig.4B), dead space is doubled so that $\dot{V}_A$ is decreased, thus elevating P_{ACO_2}. This clearly raises plant gain and consequently the post–sigh response shows a sustained oscillation. In the second case (Fig.4C), P_{ACO_2} is elevated by raising P_{ICO_2}; this also increases $\dot{V}_A$. As a result, plant gain decreases, loop gain is reduced, and the system becomes very stable. These examples demonstrate that loop gain can be altered solely by changes in plant parameters, without any change in controller gain at all. Furthermore, they indicate that hypercapnia is not necessarily a stabilizing factor and hypocapnia is not a prerequisite for the induction of PB, as has been suggested by some previous studies[18,19].

Wakefulness–sleep transitions

In the preceding discussion, we analyzed system stability assuming a steady level of wakefulness or sleep. Here we consider the dynamic events that occur when the transition from wakefulness to sleep takes place. An example of such a transition is given in Fig.5, which assumes $S_0 = 28.8$ l/min and $\tau_s = 180$ s. Breath–by–breath values of $\dot{V}_E$ (Fig.5A), and P_{ACO_2} and S_{aO_2} (Fig.5B) are shown as functions of time from the start of the transition; displayed together with $\dot{V}_E$ are the different sleep stages. Rapid withdrawal of the wakefulness drive leads to a precipitous drop in $\dot{V}_E$ and increase in P_{ACO_2}. S_{aO_2} decreases also but does so more gradually. Just prior to the occurrence of stage 1 sleep, loop gain becomes sufficiently high and the system begins to oscillate. As sleep deepens, the oscillation grows rapidly with the fluctuations in $\dot{V}_E$ attaining a maximum in early stage 2 sleep. Subsequently, with further progression of sleep state, the amplitude of the oscillation decreases until it is damped out by the time stage 4 sleep is attained. It may be noted that mean S_{aO_2}, lowered to about 93% by

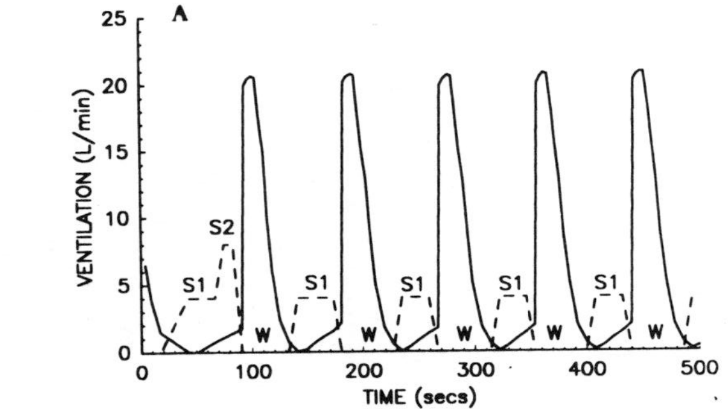

Fig. 5   Effect of wakefulness–to–sleep transition in normoxia on (A) minute ventilation, and (B) alveolar P_{CO_2} and arterial O_2 saturation. Sleep/waking states are labelled in Fig. 5A.

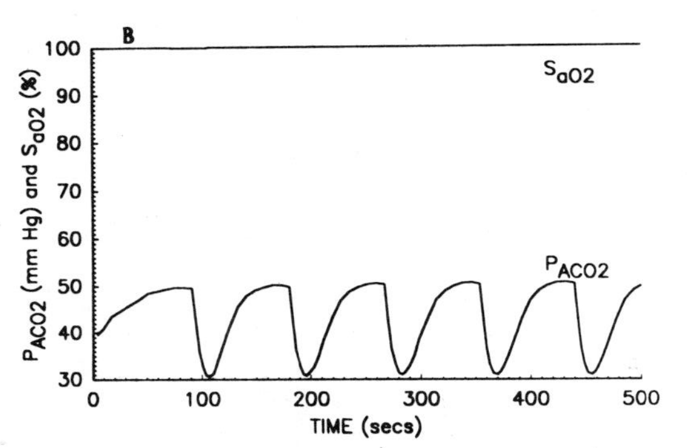

Fig. 6   Effect of wakefulness–to–sleep transition in hyperoxia on (A) minute ventilation, and (B) alveolar P_{CO_2} and arterial O_2 saturation. Sleep–waking states are labelled in Fig. 6A. After initial sleep onset, the model alternates between arousal and light sleep.

the withdrawal of wakefulness drive, recovers very slowly towards its original level during the transition. This period of mild hypoxia contributes to a shortening of the average oscillation cycle–time from 40 s (in Fig.3B) to about 30 s.

PB by repetitive arousal

The preceding example shows that, even in the absence of any other extraneous disturbance (e.g. a sigh), the rapid reduction in the level of wakefulness by itself constitutes a sufficiently large perturbation to initiate the occurrence of PB during the light stages of sleep. The underlying mechanism is a high loop–gain instability mediated by the peripheral chemoreflex. We consider next what would occur when loop gain is very low or if the peripheral chemoreflex is functionally impaired. This is achieved by changing the inspirate to a hyperoxic mixture, in this case, 50% O_2 in N_2. The result of such a simulation is shown in Fig.6. The rapid decline in wakefulness level leads to a concomitant reduction of $\dot{V}_E$ down to zero, and a short period of apnea follows (Fig.6A). This produces a large and rapid increase in P_{ACO_2} to about 50 torr (Fig.6B); P_{BCO_2} follows accordingly. At the beginning of stage 2 sleep, P_{BCO_2} crosses the arousal threshold level of 55 torr. On the next breath, arousal occurs and consequently, with the restoration of the wakefulness drive and the high CO_2 stimulus, ventilatory drive abruptly assumes large values and P_{ACO_2} is rapidly driven down. As sleep begins to set in again, the withdrawal of wakefulness drive, coupled with loss of CO_2 stimulus (resulting from the preceding hyperventilatory breaths), leads to an abrupt drop in $\dot{V}_E$. This again raises CO_2 levels and the previous cycle of events is repeated. The repetitive alternations between arousal and sleep onset occur with an average cycle duration of 88 s. S_{aO_2} remains at a steady level close to 100%. Thus, PB can still occur even when there is little or no contribution from the peripheral chemoreceptors. In this case, the underlying mechanism for the oscillatory behavior is quite different. This form of instability is associated with substantially larger fluctuations in ventilation and blood gases, and cycle–durations that are two to three times longer.

Effects of random phasic changes in ventilatory drive

Thus far, the results we have presented have been derived from simulations performed with deterministic equations and parameters. However, the existence of apparently random fluctuations in respiratory pattern on a breath–to–breath basis is well–documented, particularly during wakefulness and REM sleep[7,9]. It is therefore important to consider the effects of incorporating a limited degree of stochasticity through the inclusion of the random S_{phasic} term. In the results that follow, S_{phasic} is assumed to be zero–mean Gaussian white noise with a standard deviation of approximately 1.6 l/min.

Under conditions of elevated loop gain, as in the case of stage 1 sleep (Fig.7A), random variations in non–chemical drive can provoke the appearance of coherent oscillations. Careful analysis of the ventilation time–series in Fig.7A will reveal a strong oscillation with mean cycle–duration of 39 s in the breath–to–breath fluctuations. This oscillation is clearly mediated by the peripheral chemoreflex. Interaction between the random variations and the fundamental oscillation also gives rise to the appearance of low–frequency 'bursts'. For instance, in Fig.7A, one burst attains its maximum amplitude at around 230 s, and the following occurs at around 610 s. The occurrence of such bursts has been reported in studies of spontaneous breathing in normal humans conducted at sea–level and at high altitude[20−22]. On the other hand, when CO_2 sensitivity is substantially reduced, as in the case of REM sleep, no predominant oscillations appear and the ventilatory fluctuation have a broad frequency content (Fig.7B).

An example of the effect of random fluctuations on the ventilatory pattern accompanying wakefulness–sleep transition is shown in Fig.8. Here, S_{tonic} is assumed to change by the same amount and rate as the case illustrated in Fig.5. As in the previous case, the dominant feature in sleep stages 1 and 2 is the strong oscillation, which has an average cycle–duration of 33 s. This oscillation, although considerably attenuated in magnitude, persists through

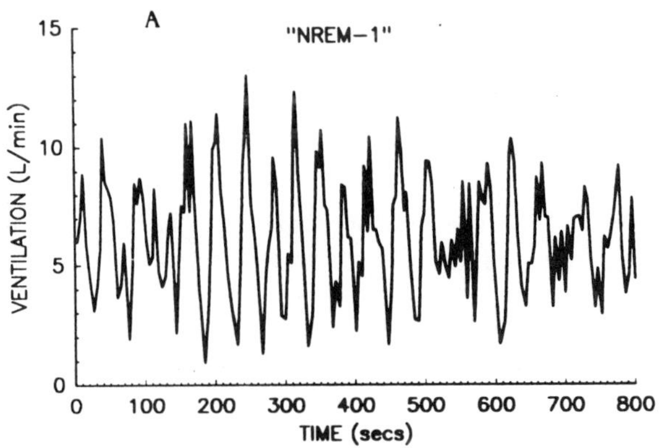
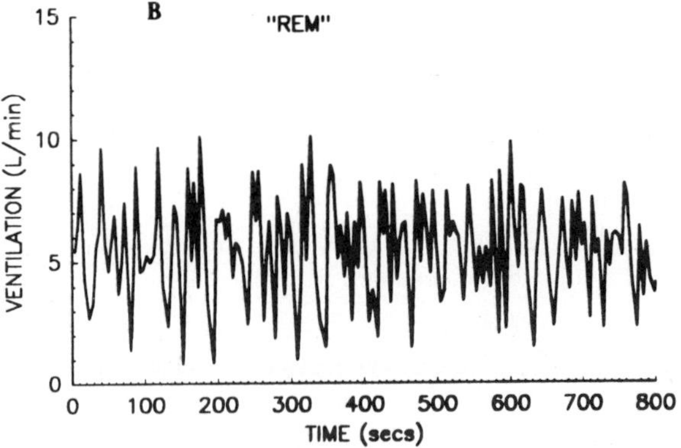

Fig. 7   Effect of random phasic variations in non–chemical drive on the model when (A) loop gain is high (e.g. NREM–1 sleep); and (B) loop gain is depressed (e.g. REM sleep).

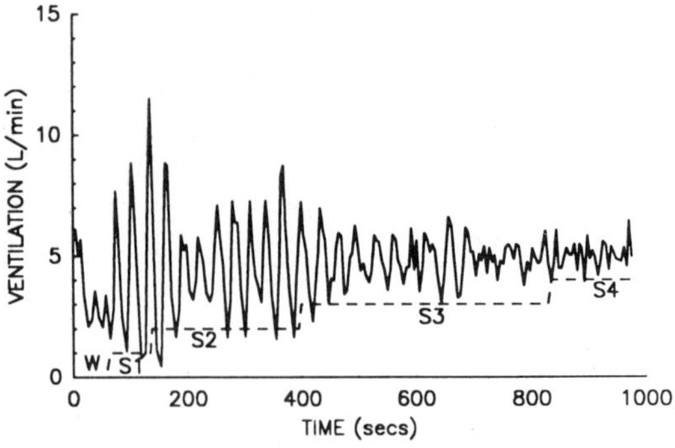

Fig. 8   Wakefulness–to–sleep transition with concurrent random breath–to–breath fluctuations in ventilatory drive.

the first half of stage 3 but subsequently disappears. As in Fig.7A, the amplitude of the oscillation is itself modulated, giving rise to the appearance of bursts: the first peaks at around 150 s, the second at around 380 s, and the third attains its maximum amplitude in the vicinity of 650 s. The bursts therefore seem to occur somewhat randomly; this feature is highly compatible with the observations of Brusil et al.[22].

MODEL LIMITATIONS

In the present model, there has been no attempt to incorporate directly the effects of changes in upper airway mechanics during sleep, although it may be argued that the ventilatory response data on which this model is based incorporates, in lumped fashion, respiratory muscle function and airway mechanics together with chemoresponsiveness. Furthermore, the state–dependent factor G_w may be interpreted as representing total airway conductance, while S_{tonic} represents the reduction in neural stimulus. However, the situation in reality is much more complex as upper airway compliance, and thus patency, can vary dynamically with a host of factors, including level of wakefulness, chemical drive and anatomical configuration[9,18,23,24]. The validity of the present model is necessarily limited to cases where no significant upper airway obstruction occurs.

The way in which we have modeled the time–course for the transition from wakefulness to sleep is highly simplistic, although it is based on the general pattern exhibited in sleep recordings. The precise points of separation between adjacent sleep states have been somewhat arbitrarily chosen. Random variations in the length of a particular sleep stage have not been taken into account. Although we have examined the effect of random variations in non–chemical drive, the effect of independent random variations in breath timing has not been considered.

We have treated REM sleep as a homogeneous state, although it has been shown that REM sleep consists of a mixture of 'tonic' and 'phasic' periods[9]. The phasic periods are characterized by eye movements, muscular twitches, and fluctuations in heart rate and blood pressure. The highly irregular breathing that is generally associated with REM sleep occurs primarily during this phase. On the other hand, respiration during tonic REM is not too different from SWS[9]. The limited data that have been collected on ventilatory response to chemical stimuli in REM sleep suggest little difference or some reduction in chemosensitivity compared to SWS. Therefore, loop gain is likely to be quite low. It is therefore unclear what mechanism can account for the occurrence of PB and apnea that have been reported in some studies on REM sleep[10,11]. One possibility is that the respiratory fluctuations and apnea which occur during REM may reflect influences originating from sources external to the chemoreflexes. This is indeed quite likely, since a control system with low loop gain is easily perturbed by extraneous inputs. Furthermore, dynamic changes in upper airway compliance, which accompany the drastic reduction in upper airway muscle activity during this sleep stage, could be an additional factor contributing to the variability in respiration.

In this study, we have assumed that 'arousal' implies the abrupt change from a given sleep state to wakefulness. Changes from one sleep state to a lighter state have not been considered, although these may be thought of as partial forms of arousal. The mechanisms responsible for arousal are still largely unknown; until these are further elucidated, the incorporation of a more complex model for arousal would involve a substantial degree of speculation.

CONCLUSIONS

In summary, based on simulations performed with a computer model of respiratory control, we have demonstrated the following points:

(1) Loop gain can be increased during light NREM sleep, if the elevation of plant gain more than offsets the concommitant reduction in controller gain; as a result, respiratory

control can become less stable at sleep onset. However, the effect of depressed controller gain predominates during SWS; consequently, breathing becomes highly regular in deep NREM sleep.

(2) The withdrawal of the wakefulness stimulus in itself may constitute a sufficiently strong perturbation to initiate an episode of sustained PB, with cycle duration ranging from 30 to 40 s. PB is more likely to occur when the magnitude of the wakefulness stimulus is greater and the rate of sleep onset is faster.

(3) In cases where loop gain is substantially depressed, regulation of blood gases can be compromised when the wakefulness stimulus is withdrawn during sleep. As a consequence, large changes in blood gases can occur, triggering arousal. The alternation between arousal and sleep onset can thus lead to the repeated cycling between hyperventilatory breaths and apnea with periods on the order of 90 s.

(4) The interaction between the random phasic variations in non–chemical drive with the chemoreflexes can lead to the random occurrence of low–frequency bursts of oscillations, when respiratory control loop gain is high. However, when loop gain is depressed, ventilation will take on the appearance of a highly irregular pattern.

ACKNOWLEDGEMENTS

This work was supported by NIH Grant RR–01861 and the Whitaker Foundation.

REFERENCES

1. E. Smith, Recherches experimentales sur la respiration, Journal Physiologie de l'Homme et des Animaux 3:506–521 (1860).

2. A. Mosso, Uber die gegenseitigen Beziehungen der Bauch– und Brustathmung, Arch. Anat. Physiol. 2:441–468 (1878).

3. A. Czerny, Beobachtungen uber den Schlaf im Kindesalter unter physiologischen Verhaltnissen, Jahrb. Kinderheilk. 33:1–28 (1892).

4. A. Mosso, Respiration on the mountains, in: "Life of Man on the High Alps", Fisher-Unwin, London (1898).

5. G. Magnussen, "Studies on the Respiration During Sleep", Lewis, London (1944).

6. E. Aserinsky and N. Kleitman, Regularly occurring periods of eye motility, and concomitant phenomena, during sleep, Science 118:273–274 (1953).

7. K. Bulow, Respiration and wakefulness in man, Acta Physiol. Scand. Suppl. 209 (1963).

8. E. Lugaresi, G. Coccagna, F. Cirignotta, P. Farneti, R. Gallassi, G. Di Donato and P. Verucchi, Breathing during sleep in man in normal and pathological conditions, in: "The Regulation of Respiration during Sleep and Anesthesia", R.S. Fitzgerald, H. Gautier and S. Lahiri, ed., Plenum, New York (1977).

9. E.A. Phillipson and G. Bowes, Control of breathing during sleep, in: "Handbook of Physiology – The Respiratory System II", A.P. Fishman, ed., Am. Physiol. Soc., Bethesda, MD (1987).

10. P. Webb, Periodic breathing during sleep, J. Appl. Physiol. 37:899–903 (1974).

11. E. Aserinsky, Periodic respiratory pattern occurring in conjunction with eye movements during sleep, Science 150:763–766 (1965).

12. M.C.K. Khoo, A model for respiratory variability during non–REM sleep, in: "Respiratory Control: Modelling Perspective", G.D. Swanson, F.S. Grodins and R.L. Hughson, eds., Plenum, N.Y. (1989).

13. M.C.K. Khoo, R.E. Kronauer, K.P. Strohl and A.S. Slutsky, Factors inducing periodic breathing in humans: a general model, J. Appl. Physiol. 53:644–659 (1982).

14. K.R. Chapman, E.N. Bruce, B. Gothe and N.S. Cherniack, Possible mechanisms of periodic breathing during sleep, J. Appl. Physiol. 64:1000–1008 (1988).

15. W.W. Pryor, Cheyne–Stokes respiration in patients with cardiac enlargement and prolonged circulation time, Circulation 4:233–238 (1951).

16. D.C. Shannon, D.W. Carley and D.H. Kelly, Periodic breathing: Quantitative analysis and clinical description, Pediatr. Pulmonol. 4:98–102 (1988).

17. B.R. Fink, Influence of cerebral activity in wakefulness on regulation of breathing, J. Appl. Physiol. 16:15–20 (1961).

18. J.A. Dempsey and J.B. Skatrud, A sleep–induced apneic threshold and its consequences, Am. Rev. Respir. Dis. 133:1163–1170 (1986).

19. N.S. Cherniack, Respiratory dysrhythmias in sleep, N. Engl. J. Med. 305:325–330 (1981).

20. J.S. Haldane, "Respiration", Yale Univ. Press, New Haven, CT (1922).

21. P.J. Brusil, "Statistical analysis of natural breathing patterns in supine man", (Ph.D. thesis) Harvard Univ., Cambridge, MA (1973).

22. P.J. Brusil, T.B. Waggener, R.E. Kronauer and P. Gulesian Jr., Methods for identifying respiratory oscillations disclose altitude effects, J. Appl. Physiol. 48:545–556 (1980).

23. G.S. Longobardo, B. Gothe, M.D. Goldman and N.S. Cherniack, Sleep apnea considered as a control system instability, Respir. Physiol. 50:311–333 (1982).

24. K.P. Strohl, N.S. Cherniack and B. Gothe, Physiologic basis of therapy for sleep apnea, Am. Rev. Respir. Dis. 134:791–802 (1986).

CONTRIBUTORS

Adrian Berkenbosch, Department of Physiology and Physiological Physics, Leiden University, Leiden 2333 AL, The Netherlands.

Eugene N. Bruce, Department of Biomedical Engineering, Case Western Reserve University, Cleveland, Ohio 44106.

David W. Carley, Department of Medicine, University of Illinois College of Medicine, Chicago, Illinois 60680.

Richard Casaburi, Division of Respiratory and Critical Care Physiology and Medicine, Harbor-UCLA Medical Center, Torrance, California 90509.

Neil S. Cherniack, Department of Medicine, Case Western Reserve University, Cleveland, Ohio 44106.

Michael Cola, Cardiovascular-Pulmonary Division, Hospital of the University of Pennsylvania, Philadelphia, Pennsylvania 19104.

Jacob DeGoede, Department of Physiology and Physiological Physics, Leiden University, Leiden 2333 AL, The Netherlands.

Jerome A. Dempsey, Department of Preventive Medicine, University of Wisconsin Center for Health Sciences, Madison, Wisconsin 53705.

Frederic L. Eldridge, Department of Physiology, University of North Carolina School of Medicine, Chapel Hill, North Carolina 27599.

Fred S. Grodins (deceased), Department of Biomedical Engineering, University of Southern California, Los Angeles, California 90089.

Adrian Goldszmidt, Cardiovascular-Pulmonary Division, Hospital of the University of Pennsylvania, Philadelphia, Pennsylvania 19104.

Allan Gottschalk, Department of Anesthesia, Hospital of the University of Pennsylvania, Philadelphia, Pennsylvania 19104.

Ronald M. Harper, Department of Anatomy, UCLA School of Medicine, Los Angeles, California 90024.

Kathe G. Henke, Department of Preventive Medicine, University of Wisconsin Center for Health Sciences, Madison, Wisconsin 53705.

Michael C.K. Khoo, Department of Biomedical Engineering, University of Southern California, Los Angeles, California 90089.

Guy S. Longobardo, Department of Medicine, Case Western Reserve University, Cleveland, Ohio 44106.

Allan I. Pack, Cardiovascular-Pulmonary Division, Hospital of the University of Pennsylvania, Philadelphia, Pennsylvania 19104.

Chi-Sang Poon, Harvard-MIT Division of Health Sciences and Technology, Massachusetts Institute of Technology, Cambridge, Massachusetts 02139.

Peter A. Robbins, University Laboratory of Physiology, Oxford University, Oxford OX1 3PT, United Kingdom.

Gary C. Sieck, Department of Biomedical Engineering, University of Southern California, Los Angeles, California 90089.

James B. Skatrud, Department of Medicine, University of Wisconsin Center for Health Sciences, Madison, Wisconsin 53705.

George D. Swanson, Department of Anesthesiology, University of Colorado Medical School, Denver, Colorado 80262; and Physical Education Department, California State University, Chico, California 95929.

Denham S. Ward, Department of Anesthesiology, UCLA School of Medicine, Los Angeles, California 90024.

Donald M. Wiberg, Department of Electrical Engineering, UCLA, Los Angeles, California 90024.

William S. Yamamoto, Department of Computer Medicine, George Washington University Medical Center, Washington, District of Columbia 20037.

Stanley M. Yamashiro, Department of Biomedical Engineering, University of Southern California, Los Angeles, California 90089.